Zero-Effort Technologies

Considerations, Challenges, and Use in Health, Wellness, and Rehabilitation, Second Edition

Synthesis Lectures on Assistive, Rehabilitative, and Health-Preserving Technologies

Editor

Ron Baecker, *University of Toronto*

Advances in medicine allow us to live longer, despite the assaults on our bodies from war, environmental damage, and natural disasters. The result is that many of us survive for years or decades with increasing difficulties in tasks such as seeing, hearing, moving, planning, remembering, and communicating.

This series provides current state-of-the-art overviews of key topics in the burgeoning field of assistive technologies. We take a broad view of this field, giving attention not only to prosthetics that compensate for impaired capabilities, but to methods for rehabilitating or restoring function, as well as protective interventions that enable individuals to be healthy for longer periods of time throughout the lifespan. Our emphasis is in the role of information and communications technologies in prosthetics, rehabilitation, and disease prevention.

Zero-Effort Technologies: Considerations, Challenges, and Use in Health, Wellness, and Rehabiliation, Second Edition
Jennifer Boger, Victoria Young, Jesse Hoey, Tizneem Jiancaro, and Alex Mihailidis
2018

Human Factors in Healthcare: A Field Guide to Continuous Improvement
Avi Parush, Debi Parush, and Roy Ilan
2017

Assistive Technology Design for Intelligence Augmentation
Stefan Carmien
2016

Body Tracking in Healthcare
Kenton O'Hara, Cecily Morrison, Abigail Sellen, Nadia Bianchi-Berthouze, and Cathy Craig
2016

Clear Speech: Technologies that Enable the Expression and Reception of Language No Access
Frank Rudzicz
2016

Designed Technologies for Healthy Aging
Claudia B. Rebola
2015

Fieldwork for Healthcare: Guidance for Investigating Human Factors in Computing Systems
Dominic Furniss, Rebecca Randell, Aisling Ann O'Kane, Svetlena Taneva, Helena Mentis, and Ann Blandford
2014

Fieldwork for Healthcare: Case Studies Investigating Human Factors in Computing Systems
Dominic Furniss, Aisling Ann O'Kane, Rebecca Randell, Svetlena Taneva, Helena Mentis, and Ann Blandford
2014

Interactive Technologies for Autism
Julie A. Kientz, Matthew S. Goodwin, Gillian R. Hayes, and Gregory D. Abowd
2013

Patient-Centered Design of Cognitive Assistive Technology for Traumatic Brain Injury Telerehabilitation
Elliot Cole
2013

Zero Effort Technologies: Considerations, Challenges, and Use in Health, Wellness, and Rehabilitation
Alex Mihailidis, Jennifer Boger, Jesse Hoey, and Tizneem Jiancaro
2011

Design and the Digital Divide: Insights from 40 Years in Computer Support for Older and Disabled People
Alan F. Newell
2011

Zero-Effort Technologies: Considerations, Challenges, and Use in Health, Wellness, and Rehabilitation, Second Edition
Jennifer Boger, Victoria Young, Jesse Hoey, Tizneem Jiancaro, and Alex Mihailidis

ISBN: 978-3-031-00475-9   Paperback
ISBN: 978-3-031-01603-5   Ebook
ISBN: 978-3-031-00037-9   Hardcover

DOI 10.1007/978-3-031-01603-5

A Publication in the Springer series
SYNTHESIS LECTURES ON ASSISTIVE, REHABILITATIVE, AND HEALTH-PRESERVING TECHNOLOGIES #12

Series Editor: Ronald M. Baecker, University of Toronto

Series ISSN: 2162-7258 Print   2162-7266 Electronic

Zero-Effort Technologies

Considerations, Challenges, and Use in Health, Wellness, and Rehabilitation, Second Edition

Jennifer Boger
Department of Systems Design Engineering University of Waterloo; Schlegel-UW Research Institute for Aging

Victoria Young
University Health Network—Toronto Rehabilitation Institute; Department of Physical Therapy, University of Toronto

Jesse Hoey
School of Computer Science, University of Waterloo

Tizneem Jiancaro
Rehabilitation Sciences Institute/Institute of Biomaterials & Biomedical Engineering, University of Toronto

Alex Mihailidis
Department of Occupational Science & Occupational Therapy/Institute of Biomaterials & Biomedical Engineering, University of Toronto, University Health Network—Toronto Rehabilitation Institute

SYNTHESIS LECTURES ON ASSISTIVE, REHABILITATIVE, AND HEALTH-PRESERVING TECHNOLOGIES #12

ABSTRACT

This book introduces zero-effort technologies (ZETs), an emerging class of technologies that require little or no effort from the people who use them. ZETs use advanced computing techniques, such as computer vision, sensor fusion, decision-making and planning, machine learning, and the Internet of Things to autonomously perform the collection, analysis, and application of data about the user and/or his/her context. This book begins with an overview of ZETs, then presents concepts related to their development, including pervasive intelligent technologies and environments, design principles, and considerations regarding use. The book discusses select examples of the latest in ZET development before concluding with thoughts regarding future directions of the field.

KEYWORDS

zero-effort technologies (ZETs), pervasive computing, artificial intelligence, disability, health, rehabilitation, wellness

Contents

Preface xi

Acknowledgements xiii

1   Overview 1

2   Introduction to Zero-Effort Technologies 3
    2.1   What is a ZET? 3
    2.2   Overview of Pervasive Computing 7
          2.2.1   Pervasive Computing Principles 8
          2.2.2   Elements of Pervasive Computing Systems 11
          2.2.3   Security and Privacy in Pervasive Computing 14
    2.3   Overview of Technology Principles 15
          2.3.1   Commonly Used Sensing Techniques 16
          2.3.2   Commonly Used Machine Learning Approaches 19
          2.3.3   Modeling Interactions between Users and ZETs 25

3   Designing ZETs 33
    3.1   Users as Collaborators 33
    3.2   Common Design Paradigms 34
          3.2.1   Universal Design 35
          3.2.2   User-centred Design 37
          3.2.3   Empathy-based Design 40
          3.2.4   Incorporating Privacy in the Design Process 41
    3.3   Ethical Considerations Regarding ZETs 46
    3.4   Key Design Criteria for ZETs 47
          3.4.1   Develop for Real-world Contexts 48
          3.4.2   Complement Existing Abilities 48
          3.4.3   Use Appropriate and Intuitive Interfaces 49
          3.4.4   Encourage Users' Interaction with Their Environment 50
          3.4.5   Support Caregivers 51
          3.4.6   Complement Each Individual's Capabilities and Needs 51
          3.4.7   Protect Users' Privacy and Enable Control over Preferences 52
          3.4.8   Ensure Expandability and Compatibility 52

4	**Building and Evaluating ZETs**		**53**
	4.1	In Silico Testing	53
	4.2	Benchtop Trials	54
	4.3	Actor Simulations	55
	4.4	Real-world Trials	56
5	**Examples of ZETs**		**59**
	5.1	Areas of Application	59
	5.2	Overview and Comparison of Examples	60
	5.3	The Detect Agitation Aggression Dementia (DAAD) Multi-Modal Sensing Network	65
	5.4	Ambient Vital Signs (AVS) Monitoring	67
	5.5	Non-contact, Vision-based, Cardiopulmonary Monitoring (VCPM) System	70
	5.6	The COACH	72
	5.7	Culinary Assistant and Environmental Safety System	75
	5.8	PEAT	78
	5.9	Braze™ Obstacle Detection Systems	81
	5.10	The HELPER	84
	5.11	Ambient Activity Technologies	86
	5.12	Autonomous Assistive Robots	89
6	**Challenges and Future Directions**		**95**
	6.1	Challenges to the Development of ZETs	95
	6.2	Future Challenges and Considerations	97
7	**Conclusions**		**99**
	Bibliography		**101**
	Author Biographies		**117**

Preface

In the 1980s, a new field of research in the area of technology and health emerged, with the aims of mitigating the effects of disease and improving rehabilitation outcomes. In the 1990s, assistive technologies (ATs) were defined as a specific class of artefacts intended to improve function for people with disabilities. More recently, the terminology has expanded beyond ATs to include quality of life and ambient assistive living (AAL) technologies, signaling the rise of a new wave of sensor-based and intelligent systems. Accordingly, over the years, technologies have ranged from simple reminding devices to help with scheduling and medication management to advanced robotic systems to support motor, sensory, and cognitive function. Regardless of the terminology, the objective has remained the same: to help people achieve their goals.

Effective, appropriate, and accepted technology solutions should offer support without increasing the workload for people living with disabilities, caregivers, and clinicians. However, using these technologies has often increased the effort required by users or transferred the effort to someone else, such as a caregiver. As a result, many devices that are intended to help are not adopted or soon abandoned by their users. In response, researchers from various fields have recognized the need for usable devices, developed with direct input from users throughout the design process; and for advanced techniques, reducing explicit user interactions.

Fields such as computer science and biomedical engineering are expanding rapidly and with them the bounds of what can be achieved with new technologies for healthcare and rehabilitation continues to grow. Incorporating advances such as artificial intelligence is greatly reducing—and in some instances eliminating—the amount of effort required to operate the technology. This book explores the exciting emerging field of zero-effort technologies (ZETs); technologies that leverage advances across multiple disciplines and sectors to provide seamless and effortless support to the people who are using them. It is through technologies such as ZETs that we gain a glimpse the future of pervasive, personalised, and holistic support for health and wellbeing.

Acknowledgements

Parts of the materials presented in this book have been reproduced from a chapter previously written by the authors with permission from IOS Press and the associated editors.

CHAPTER 1

Overview

The intersections and collaborations between the fields of computer science, biomedical engineering, and assistive technology are rapidly expanding, causing dramatic increases in the capabilities of intelligent systems for supporting healthcare and rehabilitation. As the applications of related devices become more pervasive and complex, it is important that these new technologies are designed in a way that best serves the people that will use and rely on them. This becomes even more important when these technologies are intended for people with one or more disabilities, such as people with mobility, sensory, or cognitive impairments. The special and often specific needs of people from these populations mean that they require appropriate considerations throughout the design process as the technologies may be relied upon to support a person's wellbeing. This is especially true in the development of advanced artificially intelligent technologies for supporting health and rehabilitation applications as these systems are able to perform some level of autonomous decision making while ensuring that they operate in a highly reliable fashion. Typically, these additional requirements are addressed through high levels of involvement of representative end users in the development, training, and usage of such systems. However, this is not always a reasonable expectation of all populations, especially those with complex or special needs.

This book introduces the concept of zero-effort technologies (ZETs). ZETs are technologies that operate with little or no perceived effort from the people they support; the user does not need to "learn" or remember how to use the device. ZETs often use advanced computer-based techniques, such as computer vision, sensor fusion, machine learning (ML), and the Internet of Things to autonomously operate through the collection, analysis, and application of data. This capability enables ZETs to autonomously learn about users and their environments, enabling them to support a wide variety of activities related to health, wellbeing, and quality of life. Thus, while the task the person is engaging in may require effort (e.g., performing a rehabilitation exercise), operation of the technology itself including learning how to operate it, does not. While ZETs are useful and usable by all populations, they are especially appropriate for people who are not necessarily able to reliably or explicitly provide input, as is required by many conventional technologies.

This book begins with an overview of approaches and techniques used to design ZETs for health and rehabilitation, including a discussion of design principles that can be used in the development of effective ZETs. These principles are based on established design paradigms such as user-centred design and on the experiences of the authors that include: (1) understanding and incorporating real-world contexts; (2) complementing users' abilities; (3) the use of appropriate and intuitive interfaces; (4) encouraging user involvement and participation; (5) supporting the caregiver; (6) individualisation; (7) ensuring privacy and control; and (8) expandability and compatibility with other devices and technologies. The relevance of each of these principles to ZETs is discussed, including techniques that can be used to incorporate these principles and collect the necessary data to develop adaptive systems. This discussion is followed by the ZET examples and a discussion of how design principles have been applied to improve the safety and wellbeing of the people who rely on them. The book concludes with a summary of key limitations and future opportunities for ZETs.

CHAPTER 2

Introduction to Zero-Effort Technologies

The scope and cabilities of computer science and electronics are expanding rapidly, resulting in the incorporation of technology into virtually all aspects of our lives. The purpose of technology is to enable people to do things they could not otherwise do or to complete a task with greater efficiency, effectiveness, and safety. Technology enhances a person's abilities by providing support in completing tasks and activities. In order to be effective, technologies must reflect the context in which they are being used. This has traditionally been achieved through direct user input, with the person or people using the device providing explicit and specific input, however, this requirement can often result in user burden and frustration, particularly if these inputs are needed at specific times or if a missed input results in poor device performance. For this reason, technology developers make an effort to optimally balance a device's applicability, customisability, and effectiveness with the technology's operation requirements. A goal of technology design is to minimise the number of user interactions and required effort to use it, so that there are as few demands on the user as possible. This makes technologies more useful and appealing, which will bolster adoption and use.

The ability of technologies to capture, analyse, and share data makes them a powerful tool in supporting people's health and wellbeing. It is especially important to consider the capabilities and needs of users when developing technologies intended for people living with disabilities, in particular, disabilities (e.g., cognitive, physical, sensory) that may limit a person's capacity to access and operate systems. People with impairments are often unable to reliably initiate technology use and may not possess the skills, such as procedural memory, required to operate them or learn or remember how to operate them. Such situations often result in the operational requirements of potentially helpful technologies becoming barriers to their use.

2.1 WHAT IS A ZET?

Zero-effort technologies (ZETs) are a class of technologies that operate and provide support with little or no perceived extra physical or mental effort by the people who are using them. Extra effort, in this case, would be the effort required in addition to the effort typically expended in carrying out a task or function in the absence of the technology. ZETs often employ techniques such as artificial intelligence and unobtrusive sensors to support the autonomous collection, analysis, and application of data about people and their context. ZETs operate with minimal or no explicit feedback from their users, which can include a person living with a chronic health condition, care providers, and others. This translates into minimal or no learning requirements for all the people who use them. In other words, ZETs are designed so that the people who use them can go about their daily lives as they normally would and the ZET will support them appropriately. As such, ZETs have tremendous potential to provide support to people living with chronic health conditions.

A crucial point: ZETs do not necessarily make the target activity or task effortless; *it is the operation of the ZET that is effortless*. A ZET is perpetually ready for use, requiring no perceived effort for its operation. This enables people using a ZET to centre their attention on the tasks they wish to achieve rather than operating technology. In other words, a person can use a ZET to engage in a task without focusing on operating the ZET itself to do so.

For example, smart home systems (i.e., homes that contain automated or semi-automated technologies such as heating, lights, security cameras, etc.) have the potential to incorporate a combination of ZETs that will support a person's health and wellness. As illustrated in Figure 2.1, ideally such an environment would be able to unobtrusively monitor the actions of the occupants as they go about their normal routines, autonomously learning their preferences, abilities, changes in health, etc. The system then uses this information to provide appropriate assistance only as needed. This assistance may include detecting it is time for a user to take medication and providing an appropriate prompt (e.g., auditory or visual) to do so. In this scenario, actions by the system may include detecting the user is about to go out for a walk and determining whether he should be prompted to take the required medications before leaving. The system may also detect that a user, who has dementia, is attempting to complete a self-care activity, such as handwashing, but cannot remember how to complete the activity. In response, the smart home provides a verbal reminder and ensures that the activity has been completed successfully. Finally, the system may have the ability to monitor the overall actions and movements of the occupants, learning about their daily living patterns, such as how often someone uses the washroom or when the person typically goes to bed. If changes in these usual patterns are detected, such as the person using the washroom twice as often, the home system can then automatically call for assistance and/or check in on the occupant. These different scenarios operate on the same key principle: that there is no perceived effort required from the user to operate or interact with the various devices and components. The occupant lives in his/her natural way and the technology adapts itself to his/her daily life.

Figure 2.1: A typical example of a ZET is a smart home in which the system provides appropriate assistance no matter the context or activity the occupant is trying to complete.

The success of ZETs hinges on how well they can appropriately complement the different capabilities, expectations, and goals of the people using them so that there is no *perceived* effort. For example, take the previous example of a device built to monitor a person's day-to-day routines in the home and detect when there is deviation. For the person being monitored, the device would be zero-effort if he or she does not have to do anything for the device to operate; he or she would not have to activate the system, set or change system parameters, or otherwise interact with the device for it to work. For the person's clinician, who is interested in viewing the occupant's health trends, the device could automatically generate suitable summaries, identify points that may be of interest, and raise an alert if a potentially significant situation is detected. This data would be presented to the clinician in an intuitive format that fits her workflow and could be easily manipulated, such as the ability for the clinician to drill-down to gain a more detailed understanding of the person's health, or drill-up to see a broader, more holistic view. Similarly, trends in health could be presented to the person who was being monitored and/or his family, enabling proactive health management. These data may not be in the same format as those for clinicians but will be intuitive to understand and reflect metrics or events that are meaningful for the person viewing it. This holds true for any other stakeholders who are permitted access to the data. In addition, it is not a requirement that the person being monitored and their family view or interact with the data if they do not want to do so, rather it should be an option for users who wish to do so. Thus, ZETs should be sensitive to and accommodate the variety of needs and abilities of the various stakeholders, enabling effortless access (or the avoidance thereof) to the items and functions relevant to each person's interests and abilities.

In order to implement many of the ideals of ZETs, various upcoming areas of computer science are providing the intelligence, flexibility, and comprehensiveness needed by this new class of technologies. We discuss three of these areas. First, a primary field of interest to the development of ZETs is *pervasive computing*. The word "pervasive" means "existing everywhere," hence, pervasive computing is the incorporation of microprocessors and sensors in everyday objects and environments so they can seamlessly and efficiently capture and communicate information. Pervasive computing is an underlying premise of the Internet of Things, a term that describes the concept whereby the devices are connected to the internet and each other, so are constantly available for applications that can share data across platforms. The Internet of Things enables direct and customised responses to users, environments, and geographic locations with a greatly reduced need for explicit human guidance [1].

Pervasive computing relies on the convergence of wireless technologies, advanced electronics, and networking. Leveraging the Internet of Things, data generated by ZETs can be made available in any place and at any time (e.g., health data collected about a patient can be easily accessed by the user's family physician). Security measures must be put in place to protect users' privacy and ensure that only intended recipients have access to the specific data they have been granted access to. Principles surrounding pervasive computing are discussed in detail in the following section.

A second area of interest to ZETs and pervasive computing is the concept of *context-awareness*. Context can be broadly defined as any information that helps to define the needs, preferences, abilities, and the circumstances that form the setting in which a person may participate in a particular activity. Contextual factors, as will be explained in more detail later in this book, can be defined as personal (e.g., age, gender, type of disability) and environmental (e.g., type of environment, lighting conditions, other people sharing the space) [2]. Context aware computing attempts to collect all of the data required for the developed system to provide a more customized approach in its operation [3]. This premise is critically important in the application of pervasive computing as it is these data that allow ZETs to have the adaptability and intelligence required to reduce the perceived effort of operation of their users. While the concept of context awareness is embedded in the principles that are discussed later in this book, for details specifically on this field of work and relevant research the reader is referred to more in-depth literature, such as [4–6].

Third, in order to implement many of the principles of pervasive computing and context-awareness, concepts from *artificial intelligence* (AI) are of central importance to ZETs. AI techniques can be used to manipulate and interpret incoming data, making it possible for ZETs to identify and respond to situations or trends of interest in an appropriate fashion. As such, ZETs can perform autonomous data capture, manipulation, and communication, which allows the targeted and customisable presentation of data in formats that are tailored to the interests of different people who may use a ZET, such as clinicians, professional and informal caregivers, and people with special needs.

While ZETs offer exciting potential to provide a diverse range of support, it is vital that the applications are aligned with technology's state of readiness and capabilities. The benefit of the ZET must be carefully weighed against the risks and consequences of misinterpretation of data, malfunction, or failure. Moreover, there should be adequate evidence that a ZET is the appropriate solution for each individual that they support. For example, if a lower-tech (and often cheaper) solution produces comparable outcomes, there is a good chance that it will be the preferred alternative. These concepts are discussed in more detail later in this book.

2.2   OVERVIEW OF PERVASIVE COMPUTING

Pervasive computing is an effort to make information-centric tasks simple, mobile, and secure [7]. Pervasive computing devices are not personal computers, but are either mobile or embedded in almost any type of object imaginable, including cars, tools, appliances, clothing and various consumer goods—all communicating through interconnected networks. Research in this field has primarily been aimed at improving the efficiency of current and new computing applications by making them available in more places, more often, and with more convenience for users. This concept is the backbone of the Internet of Things; a (globally) interconnected network of devices, data, systems, and services that can share data. Coupled with information such as location and preferences, the Internet of Things is rapidly growing to support applications in virtually every domain, such as manufacturing, environmental monitoring, agriculture, transportation, and healthcare. The ability of the Internet of Things to provide immense flexibility while simultaneously reducing user demands has resulted in a relatively quick uptake and implementation of pervasive computing applications, including ZETs for supporting health, rehabilitation, and wellbeing.

While application in healthcare is still a relatively new area of research in pervasive computing, this concept has already been adopted in other areas. One trend relates to technologies that support energy management and specifically "smart grid" systems [8], which is presented here to illustrate how the principles of pervasive computing may apply in practice. The IEEE's conceptual model [9] depicts the multiple and bi-directional energy and information channels between stakeholders, hinting at the flexibility and interrelated nature of this system (see Figure 2.2). This example will be used in the proceeding sub-sections to illustrate various concepts related to pervasive computing, primarily the four principles used in this field.

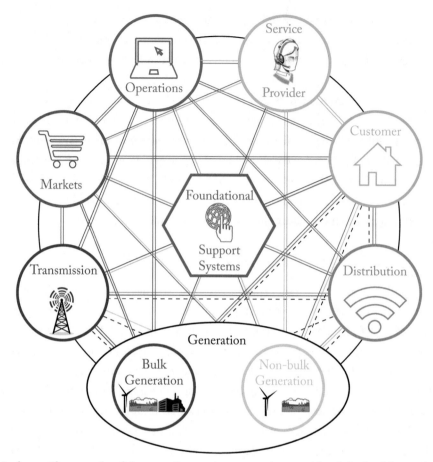

Figure 2.2: A specific example of the use of pervasive computing outside of the healthcare domain. A conceptual model of "smart grids" [9], involving technologies that are intended to be distributed, diverse, interconnected and, for the customer, simple to use (based on the Smart Grid, © IEEE).

2.2.1 PERVASIVE COMPUTING PRINCIPLES

As technologies in general are becoming increasingly pervasive and interconnected, it is worth examining the principles that guide pervasive computing. Pervasive computing is based on four fundamental paradigms: (1) decentralisation; (2) diversification; (3) connectivity; and (4) simplicity [10].

The first principle of pervasive computing is *decentralisation*. Typically, computing systems have been based on a centralised component, such as a computer server, which manages and runs all of the required computer processes. In contrast to this traditional approach, decentralisation is when a system uses many devices that that each perform a specialized task or tasks. In this way, pervasive computing distributes a system's responsibilities between a variety of devices or sub-systems, each of which take over specific tasks and functionality [10]. As each sub-system can perform some measure of independent data collection and analysis, decentralisation enables large amounts of data to be captured as each sub-system is able to perform some measure of pre-processing on the data it captures. This reduces or eliminates hardware bottlenecks and increases the transfer rate of data between sub-systems, as pre-processing allows filtering and logic operations to be applied to data pre-transmission as well as enabling techniques such as event-driven communication to be exploited. The use of cloud-based data processing and storage is another example of decentralisation and has the advantages of being accessible and has geographic redundancy, helping to protect against data loss. In general, decentralisation increases a system's robustness as it may still be able to operate effectively even if one or more sub-systems fail.

To explore these concepts, we will use an example that is relatively familiar and established: smart power grids. The notion of decentralisation is well entrenched within smart grid designs. Conventional power systems typically involve a large power plant that distributes energy downstream to end-users who possess no real-time information on rates or usage. Modernisation of power technology, however, is significantly shifting this paradigm [11]. Multiple sources of energy, including renewable energies, are now being used as well as applications that offer customers real-time information to make energy choices [12]. Opt-in plans for pervasive, non-intrusive circuitry to learn and control appliance use are being encouraged [13]. Thus, home area networks may be coordinated with ubiquitous sensor networks and operations-level networks to monitor, analyze, control, and troubleshoot power delivery [9]. In addition, there is the recent trend toward consumer-produced power, such as the use of geothermal systems to reduce heating and cooling requirements and the use of solar panels or wind turbines to generate power on-site. This approach not only provides redundancy (if the regional power plant fails, consumers generating their own power will be significantly less affected) but can reverse the power delivery paradigm, with consumers feeding power into the grid rather than drawing from it.

The second principle, *diversification*, builds upon the notion of decentralisation. In typical computing applications, a system performs all necessary applications on a centralised computer. Diversification, on the other hand, is the ability to employ several different types of devices and computing units for each sub-system, where these devices may be different "classes" of technologies, each with its own unique capabilities, inputs, and outputs. For example, a pervasive computing system can employ smart phones, laptops, and various sensors where all of these devices are able to communicate with each other and the overall system using common communication protocols (some of which are discussed later in this book). Having diversity in devices that can be used is important because it enables sub-systems to be optimised for the contexts of specific situations and environments [10]. A key aspect to ensuring that diversified systems operate effectively and efficiently is to manage the resulting diversity of data that often accompanies the use of multiple different types of devices. Data management can be particularly challenging when integrating data from different user interfaces and data input mechanisms, which can include autonomously, semi-autonomously, and manually input data captured using different computing platforms and operating systems. Interoperability between sub-systems is accomplished through the use of standards and protocols that enable different devices to perform common features and functions, which are described in more detail later in the book.

In the case of smart power grids, not only are different forms of energy increasingly being used, but also different kinds of information processing devices: home area networks are populated with smart appliances and smart meters, while non-intrusive load control circuitry functions, in part, via networks of sensors that feed into artificially intelligent algorithms [13]. These various kinds of devices meet specific requirements within the broader system, thereby simultaneously diversifying and optimising the grid.

The third principle, *connectivity*, is that all the devices within a network or system of interest are able to exchange relevant data. While the premise of pervasive computing is to allow for the decentralisation and diversity in the devices and algorithms that are implemented, there still needs to be a way for devices to communicate. This is true not only for the sub-systems within a single pervasive computing system, but for communication between these systems as well. For example, interoperability requires that the same communication protocols are used by pervasive computing devices and services within a network if they are to be able to transfer data between systems, such as data sharing between home monitoring devices and a person's electronic medical record. Just as important as being able to easily share data, pervasive systems must keep the data they share secure to prevent unwanted access. As pervasive systems become more widely applied, connectivity standards continue to be developed and adopted, such as WAP, UMTS, Bluetooth, or IrDA [10], and are described in more detail later in this book.

This principle of connectivity applies to smart grids also because although energy networks function independently, they, too, must be capable of communicating seamlessly. To meet this requirement, "Smart grid interoperability standards" [14] have been catalogued by the National Institute of Science and Technology (NIST). These standards render communication within the broader power system possible. The ability for smart devices to communicate enables the concept of power optimisation at the consumer level, giving a smart home the ability to operate devices in off-peak hours and to autonomously adjust temperature and lighting depending on whether people are at home or not.

The final principle, *simplicity*, is perhaps the most important principle of pervasive computing for the development of ZETs aimed at supporting healthcare and rehabilitation. Pervasive computing has a heavy focus on issues related to human-computer interaction (HCI). One way ZETs ensure there is no or almost no learning curve is by creating devices and systems that are so intuitive people can use them without instructions on how to do so [7, 15]. Techniques to aid simplicity include *errorless learning and exploration*, which enables a user to explore an interface without concern for faults; *skeuomorphism*, which leverages metaphors from the physical world that are appropriate to a user (such as an electronic calendar that mimics a wall calendar for those comfortable with paper-based artefacts); enriched feedback to clearly communicate the result of an action (for example, previews to foreshadow a result); and the use of *affordances*, that is, opportunities for action that are *discoverable* (such as presenting hyperlinks, for example, rather than hidden menus).

In the context of ZETs, simplicity does not necessarily mean that the technology itself is simple, rather that the *perceived* interaction with the technology is simple. In other words, while the technology may employ cutting-edge ML, sensors, and communication protocols operating as part of an interconnected global network, people find it effortless to use. How to accomplish this depends on the application, the hardware and software involved, as well as the person who is using it. Some interfaces are more conventional, such as applications that operate on personal or laptop computers that use a keyboard and mouse interface. However, pervasive computing devices are increasingly relying on sensors embedded into an environment or are small mobile devices. This changing form factor towards including sensors (and therefore input modality) is driving the development of new approaches and techniques that allow users to easily interface and interact with devices in non-traditional ways. For example, control of a device using speech is a user interface that may be more intuitive for many people. In some instances, explicit user input may be eliminated completely as AI increasingly takes on tasks that previously required human input.

The principle of simplicity is critical when developing technologies for people with disabilities or older adults. These groups tend to have differing physical, cognitive, or sensory abilities and thus greatly benefit when they are able to interact with technologies in a way that is simple, intuitive, and requires no learning.

2.2.2   ELEMENTS OF PERVASIVE COMPUTING SYSTEMS

For a pervasive computing system to be effective, there needs to be three different components that are taken into consideration: (1) the devices that run developed applications; (2) the protocols and standards that allow devices and applications to communicate; and (3) the applications and services provided by pervasive computing [15].

Devices

There is a wide range of different *devices* that can be part of a pervasive computing system. Any device that is considered to be useful to the system can be employed and can range from more novel devices, such as implanted sensors, to more "traditional" computing devices such as desktop computers and data servers. The latter devices often provide behind the scenes support for the mobile components of a pervasive application because of their more extensive computing and communication capabilities, however, they also require other devices such as network routers and modems, mobile phone towers, and wireless access points to be included in the system [15]. With respect to the previous smart grid example, devices that could be used in a person's house include sensors that detect environmental conditions within the home, such as temperature sensors in several rooms that wirelessly communicate with a networked thermostat and lighting that adapts to patterns of living, as well as devices that do not have "smart" capabilities, such as kettles, dishwashers, laundry machines, air conditioners, and other appliances. While the owner can manipulate the smart devices explicitly, smart homes are increasingly able to learn people's living habits (such as power consumption) and autonomously adapt their use to optimise costs related to power consumption within the occupants' comfort zone. The ability of a smart home to self-configure is increasing as the percentage of appliances equipped with the ability to be networked is growing.

Many advanced pervasive computing systems, including a number of applications related to healthcare and rehabilitation, rely heavily on sensors that can monitor various aspects of an environment. Examples of sensors include motion detectors, radio-frequency identification (RFID) devices, video cameras, and ultrasound. Using data from these sensors, systems can act upon the environment and/or a user through the use of another set of devices, known as actuators. Some examples of actuators are switches that can autonomously turn off a stove or the lights, or can send an alert to a person's mobile device. A more detailed discussion of sensors and actuators is presented in Section 2.2.

Standards and Protocols

In order to support the pervasive computing principle of diversification, *standards and protocols* are essential to allow different devices and systems to share information. At the hardware level, there are many standards and protocols that specify the design of physical components to ensure operability, such as standardised connection ports, ergonomic design, and safety requirements. Coupled with the hardware, software standards and protocols are responsible for how devices collect and output data to ensure that all other devices can understand the messages the device transmits or needs to receive [15]. Protocols are used to define how to package information, address messages to different recipients, to ensure messages or data have arrived at their intended destinations, and to ensure that data is not lost or compromised during transmission. Standards and protocols at both the hardware and software levels guide how to build adequate security into a system. This includes ensuring that a user or a device has permission to access a resource and preventing non-authorised users and systems from intercepting or "listening" to messages [15]. Issues regarding security in pervasive computing and ZETs are presented later in this book.

There are a multitude of standards and protocols at both the hardware and software levels. Using protocols, device manufacturers can describe the kinds of information that their hardware produces and accepts. This information enables different devices to be readily integrated into heterogeneous networks because a new or modified device essentially tells the network what data it can provide and what data it is looking for. With these protocols in place, a person simply adds a new device to an existing network, which automatically identifies the device and incorporates it into the network appropriately. This type of automatic, "plug and play" network configuration is an example of a goal of the ZET paradigm, as it allows devices to be seamlessly integrated into existing systems whenever and wherever they are needed without the user having to configure the device [15]. For example, while they do require some configuration, with smart home systems (e.g., Google Home and Apple HomeKit), the consumer selects the sensors they wish to use and the system integrates each one as it is added. Smart home control over things such as temperature and lighting impacts our smart power grid example.

The ability for different devices to be automatically and effortlessly integrated into an existing architecture addresses a key area in the development of ZETs, which is to enable compatibility between different devices and systems. For example, the issue of cross compatibility is becoming very important in new smart home systems that are being developed to help people with various abilities complete a variety of tasks and activities of daily living. Many of the available technologies are proprietary, standalone systems that require a considerable effort from software developers before they are able to communicate with each other. As smart home systems increase in maturity and popularity, communication protocols developed for integration could be adopted and leveraged by assistive technologies.

Closely tied to hardware-based protocols and standards are those related to communication networks. Network protocols have existed for many years with one of the most common and well recognised being the Transmission Control Protocol/Internet Protocol (TCP/IP) [15]. TCP/IP is built into many operating systems (e.g., Windows, Linux, Mac-OS) and is the protocol used by the Internet. This protocol is one way of specifying the addresses of different devices in a network and ensuring that messages are correctly sent from one device to another across a network. Other protocols have been built on top of TCP/IP, such as Hypertext Transfer Protocol (HTTP), which is a communication protocol for transferring data on the World Wide Web. HTTP is a server-client based protocol where one system, called the client (e.g., a web browser), makes a request to another system, called the server (e.g., a web server), for particular data (e.g., files) [15]. The server interprets the client's request, and if the request is considered to be genuine, passes the relevant data back to the client.

The dramatic increase in the use of mobile devices, such as smartphones, has resulted in large and rapid increases in the development of protocols and standards related to these pervasive computing devices. This includes protocols for cellular telephone communication, localisation, and peripheral device communication (e.g., Bluetooth). As hardware improvements enable increasingly faster transfer of greater amounts of data, new protocols and standards will arise to complement new capabilities and needs.

Application Services

The primary objective of pervasive computing is to develop applications and services that can be used effectively in a variety of contexts. The purpose of developing devices, standards and protocols is to provide the infrastructure that is needed by these applications [15]. Pervasive computing covers a wide range of applications from communication (including email, voice calls, text messaging, and video conferencing) to data management and analysis, such as database access and file transfer. For example, as part of the new smart power grid systems, many consumer hydro and power providers have developed "easy-to-use" web-based portals that consumers can log into and view their current power consumptions along with ways of reducing their power usage and costs. Layered on top of this, consumers are increasingly installing smart power-management systems for their own homes. For example, intelligent thermostats with customised temperature bands that can autonomously learn the occupants' preferred temperatures for different days of the week and times of day. As well, the user has the ability to remotely program and operate the thermostat should he or she wish to do so. As more appliances become available with smart capabilities, smart connectivity is enabling them to leverage demand and response programs. These programs allow appliances to autonomously communicate with the power grid and (when possible) run themselves during periods of low demand on the grid, which saves the consumer money and benefits the power grid as a whole by optimising power consumption [16]. This concept is highly relevant to industrial applications as well. As pervasive computing capabilities and infrastructure continues to mature, more sophisticated applications and services will continue to emerge.

The same trend in progressively intelligent and autonomous applications is the case for healthcare applications as well, where new pervasive computing systems are being developed and deployed to help in the provision of care to people receiving care in hospitals, clinics, and within the home. Pervasive computing in healthcare has included applications and services aimed at reducing the cost of providing healthcare, increasing the quality of care, providing peace of mind and assistance to family caregivers, and assisting in the management of chronic conditions. Examples of such systems related to healthcare and rehabilitation is provided in Chapter 5 of this book.

2.2.3   SECURITY AND PRIVACY IN PERVASIVE COMPUTING

The premise of pervasive computing and its principles makes systems possible targets to security threats. Making information available to users in more places and at more times requires increased collection and transmission of data, which provides additional opportunities to those who wish to steal or corrupt information [10, 15]. Intimately related to security is the notion of privacy, which is an assurance that a piece of information is only accessed by permitted users and devices [10, 15]. Pervasive computing systems can be vulnerable to privacy threats as information and data typically need to be transmitted through infrastructure that is shared by other people, such as the Internet. As a result, trade-offs exist between security, privacy, convenience, and cost that need to be considered and incorporated into the design of any pervasive computing system. For example, typing in passwords or waiting for authorisation to use resources can be a hassle, but one that is reasonable and acceptable for applications that contain sensitive data.

As more devices and applications are connected to each other, this need for multiple authorisations to occur increases, and as such, the aforementioned trade-offs become even more important. These considerations can be particularly challenging to implement in healthcare technologies as overburdened healthcare workers and caregivers do not have the time to be continuously authenticating themselves, and users who have a disability may not have the capacities required to do so. However, no user will, or should, accept a system where personal details regarding themselves and the people they care for are vulnerable to security risks. As such, it is not an option to consider security in these systems as an after-thought in the design process; security and privacy need to be priorities that are incorporated at all stages in the design process of any ZET. This goal can be achieved through a framework known as *Privacy By Design*, which was developed by the Information Privacy Commission of Ontario [17]. Details of the privacy by design framework and its relevance to ZETs and other assistive technologies are presented in Chapter 3 of this book.

2.3   OVERVIEW OF TECHNOLOGY PRINCIPLES

The potential users of ZETs for supporting health and wellbeing have diverse characteristics (e.g., people with disabilities, their families, circle of care), which makes it impractical to attempt to design one solution that can fit the needs of all users. Employing a user-centred design approach (which is discussed later in this book) is a useful way to ensure that ZETs have the flexibility to enable them to meet the various needs of each individual user as much as possible. However, applying this approach can be costly, time consuming, and difficult to support on an ongoing basis. For technologies to be made accessible by other stakeholders, such as the primary user's caregivers, family members, and support group, it needs to be universally designed to consider the range of stakeholder abilities, needs, and preferences. The individual stakeholder interests in the technology may also be quite different from one another, as will be discussed later in this book. In response, there has been an increase in the number of projects in the field of assistive technology that are applying sophisticated paradigms and approaches, such as AI, to reduce, simplify, or negate explicit user interactions. AI allows the creation of technologies that can autonomously sense, learn, and adapt to individual users, lending itself well to tasks that involve learning and decision-making. AI can help researchers to design ZETs that complement the user-centred design framework by enabling the device to autonomously customise support to the individual needs of each user. *Sensing* and *ML* are two techniques that are especially useful in achieving autonomous operation, customisability, and usability in ZETs. The remainder of this section presents an overview of general concepts regarding sensing and ML and how they relate to ZETs.

2.3.1   COMMONLY USED SENSING TECHNIQUES

A critical feature of any ZET is its ability to sense the environment. This is how the ZET determines what the state of the world is (including people of interest), which is vital for determining what a user is trying to do; how he or she is interacting with his or her environment. There is no definitive type of sensor for ZETs, rather the sensor or sensor network that is used depends on the purpose, design criteria, and specific application of the ZET. Moreover, in theory, a ZET can be expanded to include other sensors—and even other ZETs or systems—as the particular implementation requires.

Sensors can be classified according to different criteria, such as whether they are active or passive (i.e., whether they actively transmit data or not), are worn on the person or installed within the environment, and are capable of wired or wireless communication. Within these broad categories there are sensor types that are more commonly used in the development of ZETs. We will briefly discuss: (1) sensors that are embedded into an environment; (2) portable or wearable sensors; and (3) vision-based sensors (i.e., cameras and computer vision). Interested readers can seek out the many survey articles for details and further references on embedded or portable sensors, such as Chen et al. [18, 19] and Szeliski [20], or Bradski [21] for computer vision.

It is important to keep in mind that all sensors require power and need to communicate in some way to transmit data. Appropriate cables must be run for wired sensors and the user must change batteries or charge wireless sensors. Sensors range in cost, which usually reflects their complexity and accuracy. Developers need to carefully consider these aspects to ensure that the sensors that are incorporated into a ZET's infrastructure are appropriate for the population, environment, and activity that are being monitored. A proposed solution could effectively address a very real problem, but it is only useful if people can afford to buy it and are able to easily use it.

Embedded Sensors

This class of sensors consists of a wide variety of sensors that are usually low-cost, commercially available, and are often meant to measure a single factor or signal. A commonly used environmental sensor in ZETs is a motion detector that simply detects the presence of motion within a specific room. There are several versions of this type of sensor, including motion sensors that use optical, acoustical, or infrared changes in the field of view to detect motion. Motion sensors can also include occupancy sensors, which integrate a timing device to measure how long a person may be within or absent from the area being monitored by the sensor.

Other commonly used environmental sensors include devices that can measure specific events, such as mechanical devices that measure water usage, thermostats that measure environmental temperature and humidity, and microswitches that can detect if a person is lying on a bed, sitting on a chair, or has opened a door. Importantly, most sensors are only able to measure one or two aspects of an environment, usually monitor a limited area, and transmit minimal, often binary, information. For example, microswitches can only detect whether the particular door or cupboard they are attached to is open or closed. They cannot give information about how far a door is open (i.e., whether a door is fully open or open just a crack) or give information about whether someone walked through the door (as opposed to just opening it) or is interacting with items in a cupboard. Additionally, a sensor must be installed for each area or item that the ZET needs to monitor, such as a microswitch for each door of interest.

Portable or Wearable Sensors

Portable or wearable sensors are designed to attach to a person or object and collect and/or transmit data without being physically connected to a network. These types of sensors are specifically intended to be mobile and are usually as unobtrusive as possible or are designed to look appealing. As they are intended to be worn or moved, the majority of portable sensors have small form factors and are lightweight. There is a wide range of portable sensors available, including accelerometers, gyroscopes, thermisters, and heart rate monitors. As with embedded sensors, the number and types of portable sensors used in a ZET is dependent on the specific application; more than one type of sensor may be used.

Radiofrequency-based sensors are a basic example of a portable sensor that have achieved widespread use. This is because of the relatively low cost and robust performance of these sensors. A common type of radiofrequency-based sensor is Radio Frequency Identification (RFID), which consists of "readers" and "tags". Readers are usually installed at fixed points in an environment and are able to sense the presence of the tags, which are worn by a person or placed on objects of interest. A communication protocol allows the reader to detect when a tag is nearby and to identify which tag it is sensing. RFID readers require power and can be hooked into a communication network such as an intranet or the Internet. RFID tags can be passive, active, or battery-assisted passive (BAP). A passive tag generally only contains the identification number of the tag and the information contained on the tag usually cannot be altered once the tag has been assembled. In a passive RFID system, the reader periodically sends out a signal, and if the passive tag is within range, the reader can detect the presence of the tag and can identify which tag it is detecting. Using powered supplied by the reader, the coiled antennae in the tag then generates a magnetic field from which power is drawn in order to send information from the tag back to the reader, such as the tag's unique ID. Passive tags must be within a few meters of a reader to be detected. One example of the use of RFIDs is in security badges, which a user must swipe across a reader, and if the tag's ID has been registered, the reader will unlock the door. An active tag uses a battery to power the tag and is able to collect and actively transmit and receive limited amounts of data stored in its memory back to a reader. These data include the identification number assigned to the tag, but can also contain more descriptive information, such as the last known location of the tag. Similar to a beacon, active systems can be used for real-time locating by actively emitting a signal to a reader at pre-set intervals. Active tags can also be "pinged" by a reader, meaning that they transmit data to a reader only upon receiving a signal from a reader to do so. Implementing this pinging technique can greatly reduce a tag's power consumption (as data transmission is relatively power intensive) and can increase network security as the tag is not continuously transmitting data. Active tags typically have transmission ranges of up to 300 ft, although this range can be less if obstructions or radio frequency noise are present. A BAP tag incorporates other sensors such as temperature, humidity, and illumination. As such, it can collect, store, and transmit much greater amounts of data, however, this requires a larger, often external battery, which is heavier and can require more frequent recharging. The reader is referred to Want (2006) [22] for a more in-depth discussion of RFID sensors and technologies.

While the particulars may be different, many portable or wearable sensors operate in a similar fashion to RFID systems. Some sensors are capable of large and rapid capture and transmission of data, others are more passive and simple in nature. However, enhanced capabilities usually have greater power needs, resulting in heavier batteries and devices that require recharging more often. Therefore, a ZET developer strives for the optimal balance between a device's data capture and transmission capabilities and the device's power requirements.

Vision-Based Sensors

The use of vision-based sensors in ZETs, such as web cameras, is becoming increasingly popular as a result of decreasing hardware costs, vast improvements in image processing algorithms with respect to computational costs and robustness, and the recognition that the rich data set that can be collected can be used for a variety of applications. With respect to the latter point, vision-based sensors can not only collect data that other sensor types cannot, but can reduce the number of sensors required to monitor an activity or environment.

Computer vision is when a computer is used to apply various image/video processing techniques to extract information from an image or series of images (video) that is then used for another type of application. Image processing techniques include object recognition, activity recognition, motion analysis, scene reconstruction, and image restoration. In *object/activity recognition*, an image is analysed to see if it contains a specific object, feature, or activity of interest. This is achieved by the system learning what the object or activity looks like based on training data that contains various features and poses that enable the system to recognise the object or activity when the system sees it. This can be a quite challenging endeavour. For example, if a system using computer vision needs to know when a person has sat in a chair, it must be able to recognise what a chair looks like, what a person looks like, and when the person is sitting in the chair. When considering the chair alone, the system must be able to recognise many styles of chairs and that a chair can look very different from different angles (i.e., if the chair is moved or rotated) and in different lighting conditions. *Motion analysis* tracks a target object, such as a person, and is typically used in ZETs to determine what a person is doing. This is achieved by algorithms that analyse sequential video frames and compare the object of interest from one frame to the next to determine if the object is moving and, if so, in what direction and by how much. A key aspect of motion analysis is the ability of the system to differentiate the object being tracked from the background and other (irrelevant) features in the environment. Introductory texts on computer vision include Szeliski (2010) [20], Shapiro and Stockman (2001) [23], and Forsyth and Ponce (2002) [24]. Duda and Hart (2000) [25] and Bishop (2006) [26] also include many computer vision applications of ML. The most well-known practical introduction to computer vision is provided by the openCV toolkit, an open-source set of libraries for a wide variety of computer vision tasks [21]. A survey of recent advances in computer vision specifically for assistive technologies is [27].

Recently, Deep Learning (DL) has revolutionized computer vision by demonstrating impressive gains on standard CV tasks like object recognition [28] and activity recognition [1]. Large, open datasets (e.g., the MPII Cooking Activities Dataset [2]) are central to supporting the development of DL algorithms and benchmarking their efficacy. However, DL is data intensive and is sensitive to adversarial samples of data, leading to significant difficulties in its application to ZETs. DL is an ongoing and very topical research area and is expected to yield solutions to outstanding ZET problems in the near future.

2.3.2   COMMONLY USED MACHINE LEARNING APPROACHES

Machine learning (ML) is an area of AI concerned with the study and creation of computer algorithms that improve automatically through exposure to data. This area of AI is extremely important in the development and usability of ZETs as even users from the same population can be quite different and a person's habits can change drastically over time. However, so far there has not been much literature on this area within the context of ZETs and other technologies for healthcare and rehabilitation, as these fields and applications are relatively new. As such, a more detailed presentation of ML and its applicability to ZETs will be provided in this section.

There are a variety of different ML techniques that have been developed and each has its strengths and weaknesses, causing different ML algorithms to be better suited to different types of problems. In all cases, the basic idea is for the system to learn a function that maps between some inputs (e.g., sensor readings in a smart home or database query results) and some outputs (e.g., identifying a human behaviour or selecting an action for a system to take). For example, if the system detects that a flow meter is running and that the level of water in a sink is very high (inputs) it means that it should shut the water off before the sink overflows (output).

ML is often achieved using a data set that represents the problem of interest. After learning is complete, the system is evaluated on a separate set of test examples, which can be a portion of the data set that was not shown to the system in the training data or a sample of similar data from other scenarios. The metric for success is the performance of the system when interpreting and reacting to the test examples; a ML algorithm is considered to be successful if it is able to map what it learned with the training data to perform successfully on the test data. A ML algorithm can perform very well on the training data, but if it has created a mapping that is too specific to the training data, it will not be able to recognise situations of interest in other contexts and will fail when it encounters other test or real-world data. This phenomenon of performing well with training data but failing in other contexts is called "overfitting".

ML algorithms can be grouped into three categories that describe how the learning is accomplished: (1) supervised learning; (2) unsupervised learning; and (3) reinforcement learning. There are a growing number of different ML techniques in each of these three learning categories, and a review of ZET literature shows that methods from all three categories are being used in the development of new technologies. These methods are briefly discussed in the remainder of this section. The particular machine learning techniques and methods are highly dependent on the nature of the problem that is being modelled as well as other parameters, such as how much training data is available and how well this data represents the intended real-world application. Therefore, software developers must work closely with other members of the research team, such as engineers and end-users, to ensure that the problem is well defined and an appropriate machine learning technique is applied.

The remainder of this subsection provides an overview of these three categories and examples of specific ML algorithms for each. Details on the ML concepts introduced in this section can be found in texts on ML [26, 29–31] and AI [32], [33]. DL is reviewed in [34].

Russell and Norvig (1995) [33] provides a very detailed overall introduction to the field of AI and ML. Poole and Mackworth (2010)[32] approach AI from a logical perspective, presenting a simple and approachable introduction to the field. Bishop (2006) [26] and MacKay (2003) [29] provide more detailed insights into ML in particular. MacKay (2003)[29] in particular investigates the Bayesian learning paradigm in great detail. Reinforcement learning is covered by Sutton (1998) [31]. Duda, Hart and Stork (2002) [25] give an excellent introduction to pattern analysis and ML, including aspects of computer vision. WEKA (www.cs.waikato.ac.nz/ml/weka) is a suite of Machine Learning tools with open-source implementations provided online, including documentation of the various approaches. WEKA is usually used as a starting point for the application of ML and/or AI techniques.

Supervised Learning

Supervised learning is one of the most common and well-known ML approaches. In supervised learning, the data set that will be used to develop an algorithm (or model) is first labelled by an expert (usually a human). The algorithm is then presented with the training data, which consist of examples that include both the inputs and optimal corresponding outputs. For example, for a system that was being developed to use computer vision to recognise when someone was sitting in a chair, the training data set might consist of a number of videos filmed from different angles of people getting into and out of different types of chairs. For each video, a human expert would identify and label objects and situations of interest; in this example, the chair, the person, and whether the person was sitting in the chair or not. From these data, the algorithm can use the labelled data to learn relationships between objects and situations of interest so that it is able to identify and map, or classify, new inputs to appropriate outputs.

Typically, data for a supervised learning approach take the form of a number of inputs that can be discrete (taking on one of a set of values) or real-valued (taking on any real number), and an output (target). For example, in a ZET application, inputs may be the values of a number of sensors in an older adult's home, while the target attribute (output) may be a category of human activity. Inputs can represent a discrete (e.g., a switch that is on or off) or continuous (e.g., a temperature) condition of the world. Output is typically discrete (e.g., an activity is being performed or not), partially because this is generally simpler to model, but the variable can be continuous when warranted by the application, such as when a ZET is using outputs that are not directly observable (e.g., estimating a person's risk of falling).

The most well-used supervised learning technique for a model that uses discrete-valued target attributes is a *decision tree* [33], [35]. With a decision tree, input attributes are used to divide the data into sets, which are used to predict the target attribute with greater accuracy. A decision tree consists of a set of nodes arranged in a tree structure, each node being associated with a *test* of a set of input attributes (usually only one). The possible results of the test are represented as *branches* emanating from the node, leading to other, similar nodes or to *leaves* of the tree, each of which represents a target attribute value. Classification of a data sample proceeds from the root of the tree by applying a test at a node, and then following the branch that corresponds to the output of the test. This process is repeated until a leaf is reached, at which point the predicted target attribute is read off. A decision tree can be learned using a simple *greedy approach*. In the greedy approach, each input attribute is evaluated by its ability to divide the input data into sets that have similar output attribute values. The input attribute and test that performs best is then chosen as the root node. This process is repeated recursively for each test result and associated training data. The choices of attributes, how to split data based on those attributes, when to choose to stop splitting data, and when to create new branches are settings that will bias the final decision tree results. Therefore, carefully assigning and assessing these settings is key to learning an effective decision tree (i.e., one that generalises and is not overfitted). The standard text on decision tree learning is described in [35].

Another very popular supervised learning approach is the *neural network*. The neural network was one of the first forms of AI to emerge and attempts to emulate human intelligence on a machine by replicating the neuron-based learning that occurs in humans. In a neural network, each input attribute is assigned to an input node in a network of neuron-like processing units. Each artificial neuron takes values from a set of input nodes and compares a weighted sum of these values to a threshold, firing an output *pulse* if the threshold is crossed. A second layer of neurons then combines the outputs from the first layer in a similar way. The outputs of the second layer are predictors of the target attribute. The neural network is trained by repeated presentations of inputs, which results in a series of outputs. A simple update rule is used at each node to compare the outputs and bring the predictions closer to the true outputs. While this technique can be robust and versatile, training a neural network is often time consuming and may not be generalisable to other applications of interest. Chapter 5 of Bishop (2006) [26] contains a good overview of neural network training algorithms. With the recent explosion in computational power and availability of data, neural networks have taken a great leap forward by constructing many-layered versions called "deep networks" [36]. These deep networks, if trained properly with sufficient data, have been shown to outperform standard ML techniques on a number of tasks. While they may be appropriate in some ZET applications, they do require significant training data and thus must be approached and used with some care at this early stage of research.

Unsupervised Learning

In many situations, it is difficult or undesirable to assign output labels to a set of training data. For example, a ZET could be used to autonomously detect changes in a person's daily routines. However, what constitutes a normal routine is drastically different from person to person; therefore, the ZET must first model (or learn) each individual's normal routines. After the normal routines are captured, new data can be compared to this normality model to see if there are any deviations and, if so, what the deviations are. Unsupervised learning tackles problems such as these, where it is not desirable or feasible to assign categories to a training set, but rather is important to learn the unique types of patterns normally present in each particular deployment [37].

The most common unsupervised approaches are statistical in nature, where a statistical model is hypothesised and its parameters are learned by a computer from a set of data. The simplest example is the mixture model, where a probability distribution is defined over the input attributes given a set of (unknown but of fixed size) labels. These labels then constitute the target (output) attributes. An algorithm such as the expectation-maximisation (EM) algorithm can then be used to learn the parameters of this probability distribution such that the likelihood of the data given the parameters is maximized. Mixture models can handle multi-dimensional, mixed continuous and discrete inputs and can learn output attributes with many values [26]. A generalisation is to learn parameters such that the probability of the parameters is maximized given the data (Bayesian learning). In this case, a prior probability distribution over the parameters is necessary to encode all prior information that is available about the model. In this case, one usually assumes that the *number* of parameters is known beforehand (e.g., the number of target output labels we expect to see). A further generalisation removes this restriction, and allows the model to also learn the *number* of output values or labels by placing a prior distribution over this number. The resulting model will be able to learn both the values of the parameters and the number (and type) of parameters. These methods are known as hierarchical Bayesian learning or nonparametric Bayesian learning methods. Such methods have been used recently in many text categorisation and machine translation problems [38] and have started to be applied to ZETs.

Unsupervised learning can be seen as a form of clustering, in which a computer searches for patterns in the data. The *clustering objective* is to create the optimum number and size of subsets of input data such that the data within each subset are all very similar (small intra-class distance) and all the data in different subsets are very dissimilar (large inter-class distance).

Neural networks can be used for clustering. The self-organising map (SOM) is a classic example of a neural network with an unsupervised training rule [39]. In a similar fashion to the supervised case, SOMs are trained by slowly adjusting the weights of each neuron. However, the adjustments cannot be made to bring the predictions in line with the true outputs (as the learning is unsupervised), therefore some other measures of success must be used. Maximizing the entropy of the target labels given the inputs is one approach that attempts to satisfy an objective similar to the one defined above for clustering. See Chapter 5 of Bishop [26] for more details on neural networks in general and Chapter 14 of [34] for Deep Network approaches to unsupervised learning.

Bayesian networks (BNs) provide a framework for modelling uncertainty that can be used for both supervised and unsupervised learning. Many of the techniques we have been discussing under these headings can be formulated as BNs, and powerful learning techniques exist to adapt these models to data. See Koller [40] for details.

Reinforcement Learning

In reinforcement learning (RL) [31], an agent (i.e., something that perceives and acts) explores the environment and receives a reward upon achieving a goal. Rewards can be positive or negative to encourage the agent to take actions that will be the most likely to result in desirable states and dissuade actions that will likely result in undesirable states. RL can be seen as a form of supervised learning in which the output (target) labels are the reward values. The difficulty with RL is that the reward is often delayed, namely rewards may only be attainable in states that are not immediately achievable therefore the agent must take several actions before gaining a positive reward, possibly risking gaining negative rewards (i.e., a negative value of reward or "punishment" associated with undesirable states) in the process. For example, a personal assistant robot may only learn that it has done a good job at the end of the day when its performance is evaluated. It will not know which specific actions led to that reward signal. The agent therefore needs to learn how to act so that, in the long run, it achieves the maximum cumulative reward. The function that tells an agent what to do in any situation is known as a *policy* [31].

There are two major types of RL algorithms: model-based and model-free. In a model-based approach, the agent assumes some parameterized model of the dynamics of the environment (and of its actions in it) and of the rewards. The agent then gathers evidence (observations) about these parameters while acting in the world. Once the model has been learned, a policy can be computed that optimises the reward function. The advantages of a model-based approach are that it is easily interpretable and prior knowledge about the domain is relatively easy to incorporate. Model-free approaches, on the other hand, do not assume a model; rather the agent must attempt to learn the policy directly from the data. The advantage of model-free approaches is that there is less bias imposed on the structure of the environment and more complex dynamics can be learned given enough data. The most commonly used model-based approach uses the Markov decision process (MDP), which is a general model of the environment in which the state is assumed to encapsulate all information necessary to predict the future (the *Markov* assumption). *Dynamic programming* is a classical search algorithm that can be used in an MDP to guide the agent towards a goal, which can be to optimise the agent's long-term reward [41]. *Q-learning* is the simplest model-free approach, and many variants of it have been proposed, mostly focussed on efficiency gains. The basic text on MDPs is Puterman [42]. Q-learning, as well as other model-free and model-based RL approaches, are covered in Sutton and Barto [31]. Deep reinforcement learning has recently been applied to very challenging computational problems such as playing Go and Atari video games [43].

Central to the problem of RL is the exploration/exploitation trade-off. An RL agent can either exploit its current knowledge of reward (e.g., it may know of a particular action that will yield a good outcome) or it can try something new by exploring a new action that it has not yet tried. The first option is safe while the second carries some risk, but may potentially yield a higher overall reward. Many RL agents use heuristic methods to trade-off exploration and exploitation. For example, an RL agent may try a random action some percentage of the time, slowly decreasing this percentage as it learns more and more about the environment. Another alternative is *optimism in the face of uncertainty* in which an agent always assumes an untested action is best. This approach is often very effective (fast) for RL but can carry more initial risk. Bayesian reinforcement learning (BRL) explicitly quantifies uncertainty over this trade-off and is the optimal method for RL, but carries a significant computational overhead [44]. Efficient methods for BRL are a topic of much current research in the field of computer science.

2.3.3   MODELING INTERACTIONS BETWEEN USERS AND ZETS

When designing new ZETs it is important to first understand the types of interactions that may occur between the system and the user(s). In addition, the system needs to take into account various aspects of these interactions and other aspects of the context within which the system will be used. These aspects are critical as they will determine how best to model the variables that are related to the user, activity, and the environment and they will direct the appropriate choices of sensing and ML approaches. The following discussion presents an overview of some of the key variables that should be considered when developing sensing and ML algorithms for ZETs.

Uncertainty

The world is full of uncertainty, and healthcare and rehabilitation are no exception. Uncertainty becomes an even larger issue when deploying technologies into the homes and communities of potential users, which are typically highly unconstrained, dynamic, and unpredictable environments. Sources of uncertainty include noise from sensors, unobservability of events and states, and uncertain effects of actions. The latter is particularly important when working with people with disabilities, whose behaviours are sometimes difficult to predict, especially when supporting people with cognitive impairments. For example, when assisting a user during a self-care task, an assistive system needs to sense what the user is doing using some combination of sensors. These sensors carry with them explicit uncertainty, which will sometimes providing false readings. Further, the technology cannot directly measure various "hidden" states such as a user's emotional state, awareness, level of frustration, or responsiveness to prompts (this is discussed more in respect to affective computing later in this book). Finally, a user's abilities and reactions to the system may not always be predictable (i.e., they may change from day-to-day or over time), even though the user may be completing the same task within the same environment. A system that accounts for this uncertainty will be able to make better decisions than one that does not. A reading from an unreliable sensor, for example, may lead the system to deploy a more reliable (but more expensive) sensor rather than taking action based on the first reading. The system must have a model of this reliability (uncertainty) if it is to make this choice.

Modelling uncertainty, however, comes at the cost of an increased complexity of the model, since it must take into account more factors. ML can play a significant role in helping to better model these various factors, thereby enabling a system to learn about a particular user, detect changes in his or her abilities, and adapt appropriately both in short- and long-term contexts.

Model complexity and uncertainty are treated at length by MacKay (2003) [29]. A general tool for modelling uncertainty is the BN, and the decision network (DN—also known as an influence diagram), and their dynamic counterparts (when time is involved), the Dynamic BN (DBN) and dynamic decision network (DDN). A MDP is a particular type of DDN. More complete treatments of BNs and DDNs can be found in Koller (2009) [40].

Time

A person with a disability often does not need help just once, but rather will need it repeatedly, through different tasks, and at different times of the day. As such, any assistance provided by a ZET needs to be an ongoing interaction, requiring the system to build and maintain an explicit model of time and a history of the user. AI-based approaches typically model time in one of two ways: event-based and clock-based. Event-based approaches use events as delineations of time. For example, a person entering the kitchen denotes the start of a kitchen event that lasts until the person leaves the kitchen. These events may have even finer resolution, such as the person touching a water faucet, indicating the start of that specific sub-step of a task, such as handwashing. Event-based modelling is very intuitive and powerful, as it allows for hierarchical modelling of nested events and corresponds to human perceptions of time, therefore may be easier for developers to define and model. Event-based modelling can also be more sensitive to different user's abilities, as it is less focused on how long it takes someone to complete a task (event), but rather it focuses on whether or not the task has been initiated/completed. Clock-based approaches are a special class of event-based modelling in which the only events are (regular) time intervals taken from a clock. In other words, the system samples the environment (sensors) and performs logic operations at predefined time intervals; the end of time interval, when data collection and operations are performed, is considered to be an event. The advantage of a clock-based approach is it removes the need to specify what an event is, as it will perform data sampling and analysis after each time interval regardless of what is occurring in the environment. A drawback of the clock-based approach is that there may be far too many (if the time interval is set too short) or too few (if the time interval is too long) events for a particular task, resulting in oversampling and analysis or missed real-time events in the environment.

Adaptivity

One of the primary reasons for using AI in the design of ZETs is to create systems that are flexible and adaptable without sacrificing predictability. Different people behave in different ways when presented with the same situation. Humans are also dynamic, and change over time; how a person reacts to a situation may change from one day to the next. This is especially true for people with disabilities, as many disabilities are progressive and cause significant changes in a person's abilities over time (both in the short and long term). Therefore, ZETs must have the ability to detect changes in a user's abilities and be able to adapt to these changes over time. However, this adaptability must be incorporated in such a way that it does not cause confusion or distrust in the technology; the ZET should operate in a predictable fashion, including when and how (autonomous or user-initiated) changes occur in its functioning.

The primary way to approach adaptivity is through *learning* and/or *inference*. Both approaches assume the model being used has some unknown parameters that govern how the model works. For example, an assistance system may model the probability that a user needs help with a specific task. Once the system knows this probability, it can better tailor its assistive actions to the user: it has adapted. The *learning approach* attempts to find an estimate for the parameters given a set of training data. This estimate is usually the one that results in the best description or explanation of the data. This can take the form of either the most likely parameter setting given the data, the most likely parameter setting given the data and some prior information, or the most probable *distribution* over parameter settings given the data. In machine learning, the first case is known as maximum likelihood (ML), the second case as maximum a-posteriori (MAP), and the third case as *Bayesian learning*. In the MAP or Bayesian approach, one needs to define a prior over the parameter or parameter distributions, encapsulating knowledge about the types of distributions that may be encountered in the population in question. The additional descriptive power of the model in a MAP/Bayesian approach usually comes at the cost of more complex learning, and is more general. Detailed analysis of both of these methods can be found in MacKay (2003) [29], Koller (2009) [40], and Bishop (2006)[26].

A related approach is to assume a discrete set of parameter values, and to characterize membership of the set with a label index (class variable). *Inference* can then be used to infer the value of this class variable, and thereby which parameter settings (from the discrete set) are best at describing the data. In the previous example, a factor that affects the probability of a user needing assistance could be cognitive impairment level (e.g., mild, moderate, or severe), or impairment type (e.g., cortical vs. sub-cortical). Some fixed parameters would govern how a user's need for assistance during task completion depends on these variables. The model can then be used to infer the person's level or type of impairment and the system can use this to change its response characteristics so that they complement the user's needs and abilities.

Abstraction

ZETs are commonly used to sense real-world events, which often result in a large quantity of data that comes from a variety of different sensors. For example, a system assisting a person to complete a self-care activity might contain large volumes of raw data captured using video, switches, and other sensors to detect the user's hand and body positions and interactions with various objects. The data may also contain extraneous data that are irrelevant to the task at hand. Furthermore, video-based data are often collected with high frequency (typically around 10–30 frames per second) to ensure the system can detect changes in the user and environment. Even with a simpler type of sensor, such as a switch, the raw data emanating from it will be sampled at a rate that could provide a large amount of data over a period of hours, days, weeks, or years. As such, ZETs need to be able to sift through all of these data, and create appropriate abstractions of it, both in time and in space, and over a range of sensors. Moreover, in a continuous application (i.e., the instalment of a ZET in an environment for an indefinite period of time), it is important that the ZET is able to compact or cull the data over time. Abstractions of classes or categories are critical in order to allow for simple and fast decision making for actions by the system.

In ML there are two primary techniques used for creating and learning appropriate abstractions: generative and discriminative. *Generative techniques* attempt to model the complete distribution over all the sensor data by learning a function that maps between the abstract categories or classes and the raw sensor data. This function typically is very large and complex and can be quite difficult to learn. *Discriminative techniques* attempt only to find a method for classifying the data into the necessary categories, without worrying about the complete distribution of the data. A simple example is a water flow impeller in a pipe to detect if a person has turned on the water. A generative approach will build a function describing the distribution of sensor readings for each situation: water on or off. The distribution might take some parametric form, such as a Gaussian, or might be described by a nonparametric form, such as a histogram of values the sensor reading normally takes on for each of the states of the water flow. A discriminative approach will find the threshold for the sensor's output that will be the best predictor of the water being on or off. It should be noted that in theory, a generative modeling approach will always outperform a discriminative one if it is a correct model (if it is expressive enough) and if enough training data are available to learn the model. In practice, discriminative techniques offer better performance in terms of classification, particularly when only limited training data are available. Bishop (2006) [26] provides a comprehensive overview of both types of approach.

Specification and Preferences

An important aspect in developing ZETs is to ensure that the system aligns as closely as possible to the preferences and needs of the user. This aspect of *specification* involves two related problems: (1)

specification of the machine-interpretable model of the interaction between the user, system, and task for which a person needs assistance; and (2) specification of a user's *preferences* over the various outcomes. Typically, the first task is approached by engineers who gather information about the task and convert it into a usable model for assistance, while the second is approached by human factors specialists who gather information from end users about the relative worth of the various outcomes (as described later in this book in Section 3.2.2).

The model specification involves the relationships between various elements of the task (including dynamics over time) and the user (their abilities, for example), and between the task and the sensors that are used to gather information from the environment. Engineers can usually gather this information and encode it in an appropriate model, but each new task requires substantial re-engineering and re-design to produce a working system. The automatic generation of such systems can substantially reduce the manual efforts necessary for creating and tailoring the systems to specific situations and environments. In general, the use of a-priori knowledge in the design of ZETs is a key unsolved research question. Researchers have looked at specifying and using ontologies [18], information from the Internet [45], logical knowledge bases [46, 47], and programming interfaces for context aware human-computer interaction [48].

The SyNdetic Assistance Process (SNAP) system [49] tackles this problem by starting with a description of a task and the environment that is relatively easy to generate. Interaction Unit (IU) analysis [50], a psychologically motivated method for transcoding interactions relevant for fulfilling a certain task, is used for obtaining a formalized, i.e., machine interpretable task description. This is then combined with a specification of the available sensors and effectors to build a working model that is capable of analyzing ongoing activities and assisting someone. The resulting model is called a SNAP. A SNAP is a Partially Observable Markov Decision Process, the specification of which is reduced to the IU task analysis complemented by the specificaiton of a few key probabilities (e.g., the probability a person will respond to a prompt if given, or the probability the person will lose awareness in a task) and utilities (numerical encodings of a user's preferences—see below). Current work on SNAP is to provide an easy-to-use online interface for secondary end-users who are familiar with the tasks a person is needing assistance with, but unfamiliar with the complexities of the AI models used to build the automated assistance system. This novel approach helps coping with a number of issues, such as validation, maintenance, structure, tool support, association with a workflow method, etc., which were identified to be critical for tools and methodologies which could support knowledge engineering for systems that plan and act in an environment.

The second aspect of specification is *preference elicitation* (PE). The key to any ZET is the encoding of the preferences of a user. These preferences may be as simple as a list of things that a person likes or does not like (e.g., he or she doesn't like audio prompts in a male voice), may be more complex (e.g., he or she likes audio prompts when in the kitchen, but not in the bathroom), or may be relational (e.g., likes audio prompts better than video prompts). Note that the specification of preferences is distinct from the model specification discussed above (e.g., using SNAP). Model specification simply defines the ways in which a user *can* or *normally will* interact with a system or environment, whereas preference specification encodes what the user really *wants* or *desires*. Preferences can therefore encode things outside the system (e.g., the user does not like any outcome suggested by the system, but perhaps has another, unknown outcome they are striving for).

While it is important for designers of ZETs to incorporate these preferences during the design phase, it is also critical for the technology itself to be able to extract this information during the operation of the system to ensure that the changing needs of the user are met. In ML, this can be achieved through the concept of *utility*, which can map a user's preferences onto a numerical utility function. Utility concepts originate in the study of game theory, operations research, and decision theory. Essentially, game theory guarantees that a rational human's preferences can be mapped onto a numeric scale of utility (i.e., the outcomes a person prefers will have a higher utility) and that decisions made according to this numeric scale will be the best, or optimal, for that person. However, it is known that humans do not always act rationally, and this may be even more apparent in a population of users with a cognitive disability. The study of PE—how to extract meaningful preference information from users—is an open research problem, but recent advances have allowed for its application in a variety of domains [51]. Recently, the fields of *imitation learning*, *apprenticeship learning*, *transfer learning*, and *inverse reinforcement learning* have come to the fore in AI research as methods for approaching this problem, as described in more detail in the Spring 2011 issue of *AI Magazine*. These methods attempt to learn user preferences by observing a teacher, or by transferring learned preferences from other tasks.

In general, specification is closely tied to learning (as described previously), as the models and the preferences can change over time, and learning must be used to adapt these elements as the users and environments change. The specification task is one which is typically seen as taking place before any deployment, but can also be interleaved with actual use (e.g., a caregiver, as secondary user, may refine the model while it is not being used, or the user may explicitly change their preferences).

Emotion

Emotion is a key element of human motivation and technologies that are able to align with end users on an emotional level are thought to hold great promise in the near future. Disability and health issues, including those associated with aging, create emotional difficulties and anxiety, sometimes leading to depression. These factors can have a significant impact on the quality of life. Recent work at developing zero-effort technologies that understands and responds to user emotion falls into the remit of *affective computing* [52]. Affective computing researchers tackle three core problems: recognition of emotion (from, e.g., video), modeling emotional interactions, and generation of understandable emotional displays. Recent work has applied techniques from affective computing in the development of systems to help persons with depression, anxiety [53], and dementia [54]. In the case of Alzheimer's, it is known that the disease can significantly disrupt a person's sense of self, and this can have a large impact on their ability to respond to ZETs. Recent work has attempted to tackle this problem by using social-psychological models of affect that focus on coherence between a person's sense of self and how they are made to feel in an interaction. This work has found that, while a person's *denotative* sense of self may be lost (they can't remember who they are, or who others are, or what the situation is), that person's *connotative* sense of self is more preserved (they can remember how they should feel) [55].

Experimental Performance

All the techniques presented thus far rely on some sort of training data. However, the actual aim of the learning process is to provide accurate classifications when the system is presented with data it has not seen before. That is, a classifier that does well on training data is not generally useful, as developers know the dataset (what the labels are, for example) and can manipulate the algorithms being used until the system is able to correctly classify this data. A recurrent problem in machine learning is that of overfitting, in which a classifier is trained on a static set of training data, on which it learns to perform very well (e.g., predict the labels), but then fails to perform well on test data. The model in such a case is too specific and usually overly complex in the sense that the classifier has modeled the training data too closely and, therefore, does not generalise well to real-world applications where the data may well be different. For example, if a vision-based system to recognise when a person is sitting in a chair is trained using only one type of chair, it will likely fail to recognise chairs that have a different design.

Avoiding overfitting is an art in itself that requires the model designer to carefully select the parameters to be learned so that they reflect the complexity and application of the classification problem appropriately. A simple technique for testing for overfitting involves separating the training data into two sets: a training set and a validation set. The ML algorithm is then applied to the training set then is evaluated on the validation set. The classifier that performs best on the validation set is the one that is most likely to generalise the best to other data. In practice, only a fraction (e.g., 10%) of the data set is needs to be reserved for validation. While testing a model using the validation method described above can be a good indicator of model generalisability, it is important to consider that the validation data may closely resemble the training data. For example, a data set for the chair recognition problem may contain images of many different types of chairs. However, if all the images were captured at the same angle, the classifier may do well on the validation data, but not perform well when implemented into a real-world environment. Therefore, developers must carefully select training data to ensure that it represents what the system would encounter in a real-world application as much as possible. A complete technical treatment of model selection and overfitting can be found in MacKay (2003) [29].

CHAPTER   3

Designing ZETs

With the above discussion about ML in mind, it is easy to see why it is crucial for ZET designers to gain a good understanding of the problem area and application parameters prior to embarking on creating the technology itself. As is discussed in the next section, understanding and defining the problem scope necessitates collaborating with people from other disciplines and sectors, such as clinicians, caregivers, and the intended end users of the system.

As noted in the previous sections, it is vital that every ZET is designed in a way that ensures it meets the needs and abilities of the people (or populations) that will be using it. The principles of pervasive computing, artificial intelligence, and ML allow for sophisticated features and functions to be implemented into a ZET, however, this does not ensure that the resulting technology will be useful or accepted. In other words, while a device may be able to correctly interpret and react to environments and users, it also must be usable and effective from the perspective of the people using it. It is a lack of usability and thus satisfaction that often results in the abandonment or outright rejection of a technology by a user. To support user acceptance and uptake, designers of ZETs need to be aware of practices and paradigms that can be used to gain an understanding of potential users and to build devices that complement their needs and abilities. Gaining a good understanding of potential users will also enable developers to collect targeted data that can then be used to build more effective and efficient technologies.

3.1   USERS AS COLLABORATORS

One of the most effective and efficient ways of accessing targeted users' needs, abilities, and contexts is to include representative end-users as core team members throughout the design, testing, and commercialisation processes. While including end-users is beneficial to product development, it is perhaps even more important for ZETs for supporting health then for other types of technologies. For example, ZETs often support with people who may have disabilities that are difficult or impossible to emulate. They also often have special considers regarding ethical use (especially with vulnerable populations), must complement care systems, and may have a higher dependence on reliable operation.

The field of human factors offers many approaches for including end-users in design. One of the most well-known and widely used is *user-centred design*, which is where users are routinely consulted regarding the design. This paradigm can be complemented by techniques such as agile development, where the project is executed in shorter timeframes with smaller-scale deliverables.

There are several ways to engage end-users in the design and development process. One of the most well known and commonly used is user-centered design, which is discussed in detail in Section 3.2.2. Another technique is *participatory design* (also known as "co-operative design" or "co-design"), where users contribute to and/or manage the design process more extensively and independently. In participatory design the representative end-user(s) are central and key team members who are considered a core developer of the resulting product or system. While participatory design can be a powerful technique, it is still relatively rare because of the required time, knowledge, and other commitments. *Contextual design* is an ethnographic method that can be used to gain insight into a technology's actual use. In contextual design, developers shadow stakeholders of interest and capture data regarding how the device or system is being used. This technique can be used for many aspects and levels of product readiness, such as refining prototypes and estimating impact on workflow.

Beyond the advantages of significant stakeholder involvement in development, research is increasingly required to clearly demonstrate patient and public involvement (PPI). INVOLVE, a government funded programme by the National Institute for Health Research, defines inclusion in research in three ways [3].

- **Involvement:** where members of the public are actively involved in research projects and in research organisations.

- **Participation:** where people take part in a research study.

- **Engagement:** where information and knowledge about research is provided and disseminated.

Including users as collaborators supports identifying and addressing of key design criteria discussed in Section 3.4, as well as better approaches to complex issues such as ethical use and information transparency.

3.2   COMMON DESIGN PARADIGMS

When designing for people with disabilities there are a variety of approaches that can be applied to ensure that the resulting technology or system is usable and appropriate for the targeted user(s). These approaches range from designing technologies to match the needs of a specific group (or type) of users who have similar abilities to designing technologies that are intended for a broad group of people with varying abilities and needs. In the development of ZETs, the two most commonly used paradigms—i.e., ideas underlying a methodology in a particular subject—are universal design and user-centred design. A more recent paradigm is empathy-based design. It is important to note that, as with most general principles and paradigms, these paradigms are concepts. As such, it is unlikely that any device can achieve full compliance because of unavoidable design conflicts. For example, situations may arise where two conflicting goals cannot be reconciled and a trade-off must be made. Therefore, designers should use these concepts as a guide to gain a deeper understanding of the problem and application; identify potential conflicts in desirable outcomes; carefully weigh trade-offs; and strive to achieve the best possible outcomes.

3.2.1   UNIVERSAL DESIGN

The goal of *universal design* (UD) is to create products and environments that can be used and experienced by people of all ages and abilities to the greatest extent possible [56]. In the context of ZETs, developers try to apply the principles of UD to create a technology that can be used by as many different people as possible with as little adaptation or learning as possible. This concept does not only apply to developing a technology that can be used across different individuals, but also to developing a technology so that it can be operated by the user and his/her caregivers, family members, etc. Often the latter is a more important application of UD to ZETs and other assistive technologies. In the best examples, UD features go unnoticed because they have been fully integrated into design solutions that are used by the full spectrum of the population, including children, older adults, abled individuals, and people with disabilities. UD also incorporates accessibility features that are recommended or required by standards, codes, and legislation; however, in the most successful applications, these features are often not noticeable as they blend into the overall design [56]. The key concept of UD is that it is not associated with the outcome of a design task, but with a process and mindset used throughout the design process.

The term "Universal Design" was first coined by Ronald Mace at North Carolina University, where he was the founding director of the Center for Universal Design. From its early definitions, UD was further developed by the Center to include seven principles as described below. The goals of these principles are to provide designers of products and environments with a set of objectives that can be easily followed and implemented into any design process, as well as, a simple validations tool that can be used to assess whether a design is accessible and usable by a wide range of individuals. While these principles were initially developed primarily for environmental design, such as architecture and landscaping, they are starting to be more widely applied to the development of physical products and computer-based systems, such as ZETs, as well as non-tangible products, such as graphic design and communications [56].

Principle One: Equitable Use

The principle of *equitable use* states that a design should be useful and marketable to people with diverse abilities. Specifically, the design needs to provide the same means of use for all users; identical whenever possible, equivalent when not. Furthermore, the design must attempt to avoid segregating or stigmatizing any user as a result of an impairment, disability, or handicap, which necessarily includes making proper and equally effective provisions for safety and security. Finally, the first principle stipulates that a design should be appealing to all users with respect to aesthetics, materials, installation, etc.

Principle Two: Flexibility in Use

The second principle, *flexibility in use*, is aimed at ensuring that a design accommodates a wide range of individual preferences and abilities. This includes providing a choice of how a user can interact with a product or environment, such as accommodating left- and right-handed access and use; facilitating users' differences in accuracy and precision; and providing adaptability to different learning styles and paces. The latter is especially important when designing systems that are specifically for users with cognitive or learning disabilities.

Principle Three: Simple and Intuitive Use

The principle of *simple and intuitive use* states that a technology or system must be easy to understand and use, regardless of the user's experience, knowledge, language skills, or cognitive abilities. This notion aligns with an aforementioned goal of pervasive computing and ZETs, which is to reduce or eliminate the learning that is required to use a new technology. Simple and intuitive use is accomplished by reducing or eliminating unnecessary complexity; being consistent with user expectations and intuition; accommodating a wide range of literacy and language skills; and providing effective prompting and feedback during and after task completion.

Principle Four: Perceptible Information

The *perceptible information* principle builds upon the concept of providing effective prompting and feedback to a user by ensuring that the design communicates necessary information effectively to the user, regardless of ambient conditions or the user's sensory and cognitive abilities. Specifically, the design or system should use different feedback modalities (e.g., pictorial, verbal, tactile, etc.); maximize the probable uptake of essential and critical information; and provide compatibility with a variety of techniques or devices that a person may already be using, such as an assistive technology.

Principle Five: Tolerance for Error

The fifth principle, *tolerance for error*, states that a design needs to minimise hazards and potential adverse outcomes that may result from accidental or unintended actions. The design should arrange elements (e.g., buttons, controls, handles, and stairs) in a way that minimises hazards and potential errors; provides clear warnings of potential hazards or errors; provides fail-safe features; and discourages unconscious actions that could be hazardous.

Principle Six: Low Physical Effort

The *low physical effort* principle states that the design must be able to be used efficiently and comfortably with a minimum amount of fatigue. This can be achieved by allowing a user to maintain neutral body postures; use reasonable operating forces; and require few or no repetitive actions. For example, if the technology in question is being designed for exercise purposes, even though the user may be fatigued from the physical exercise, the technology itself will be designed to comfortably support the user to efficiently complete his/her desired exercise task. This principle can be extended to include low cognitive effort in order to accommodate those users with cognitive or intellectual disabilities. The principle of low physical effort is at the heart of many assistive technologies and ZETs, as these devices are specifically designed to make everyday tasks easier for people, both with and without disabilities.

Principle Seven: Size and Space for Approach

The seventh and final principle, *size and space for approach*, is focussed primarily on the environment within which a person is completing a task. It states that appropriate size and space is provided for approach, reach, manipulation, and use regardless of the user's body size, posture, or mobility. This includes providing users with a clear line of sight to important elements and ensuring that adequate space is provided for the users who have assistive technologies (e.g., a wheelchair) or personal assistance (e.g., a caregiver).

3.2.2   USER-CENTERED DESIGN

Often, ZETs and other technologies for health cannot be designed so they are useable, and/or useful by the general public, since these systems perform highly specialized functions applicable to only a subset of the population. Accordingly, designers need to consider the primary users and how their particular capabilities may impact the ability to effectively and safely operate the technology under development. This approach is known as *user-centred design* (UCD), which is a broad term to describe design processes where (representative) end-users and experts in different fields influence how a design takes shape [57]. UCD aims to create technology that identifies and meets the targeted users' needs as fully as possible, which can range from a single person to the general public, depending on the application. UCD is a general methodology and is commonly used in engineering design and human-computer interaction. It has been characterized by four activities, as described by Gould and Lewis [58] and later expanded upon by Wickens [59]:

1. seeking to understand user characteristics early in the design process;

2. adopting empirical techniques such as questionnaires, interviews, observations and focus groups to do so;

3. applying an iterative design-and-test cycle; and

4. employing participatory approaches in which users are, to varying degrees, involved in the design process.

Within UCD there is a wide variety of techniques and ways in which users are involved in the development of the technology, but the important elements of this paradigm are that users are involved one way or another and that the designers actively take into account the special needs and characteristics of the targeted users. For example, user feedback will help with preference elicitation as previously described in the section on machine learning techniques.

UCD can be implemented based on a variety of approaches. For example, Norman (2002) proposed four primary recommendations to ensure that the user is always at the centre of the design [57]:

1. make it easy for the user to determine what actions are possible at any moment;

2. make things visible, including the conceptual model of the system, the alternative actions, and the results of actions;

3. make it easy for the user to evaluate the current state of the system; and

4. follow natural mappings between intentions and the required actions, between actions and the resulting effect, and between the information that is visible and the interpretation of the system state.

In a similar approach, Preece et al. (2002) [60] describe the concept of *interaction design*, which is defined as "designing interactive products to support people in their everyday and working lives". Their process involved four basic activities:

1. identifying needs and establishing requirements;

2. developing alternative designs that meet those requirements;

3. building an interactive version of the chosen design so that it can be communicated and assessed; and

4. evaluating what is being built throughout the process.

These activities are intended to inform one another and then be repeated as necessary [60].

By applying these various models and concepts the designer can facilitate a task for the user, ensuring that the user is able to make use of the product as intended, with a minimum of effort to learn how to use it. This concept is not only applicable to the final product, but also to all stages of the development and prototyping process.

The UCD process begins by discovering the users' requirements, which are used to compile the functional specifications of the device. This step often needs to be revised several times throughout the development process as the device's design evolves [61]. Discovering users' needs necessarily involves actual users and often has users performing tasks or activities in the environment(s) where they would use the technology. It is important for developers to remember that there are often several different users of a technology that go beyond the targeted main end-user. For example, a ZET that is designed to monitor and support a person with a chronic health condition may also be used directly or by proxy by his or her caregiver and clinicians. Each user type, and indeed each individual user, may have quite different opinions about what information and functionalities are beneficial as well as how these data should be presented, who should be able to see the data, and how the user should interact with the system.

There are three different types of users that need to be considered in the UCD design process: primary, secondary, and tertiary [62]. Primary users are the people who actually use the device. Secondary users are those will occasionally use the device or those who use it through an intermediary, such as a caregiver having to help a person use their particular device. Tertiary users are people who will be affected by the use of the device or make decisions about its purchase, but do not use the device themselves. The inclusion of these different types of users is especially important in healthcare as the needs and expectations of a user's family members, caregivers, clinicians, and other health practitioners may also need to be considered. This is especially true in situations where the primary user is not making the decision to purchase a specific technology or is not the one paying for the device (as is often the case with technologies that support someone who has a disability).

Once the users and other stakeholders have been identified, information about their needs and preferences need to be collected in a systematic way through methods that may be structured, unstructured, or both [61]. Gathering structured information from users can take many different forms, such as questionnaires and interviews; formal measurements of users performing a specific task; and physiological measurements. Unstructured approaches may include comments made by users during an evaluation and observations made by developers. Structured data usually provides details about physical specifications, such as dimensions for components, while the unstructured data provides insight into preferences and can identify approaches that developers may not have considered.

A key component of the UCD approach is to develop several iterative prototypes of a technology; and to evaluate them with the identified users and stakeholders as often as possible. The number of prototypes that will be needed depends on the project and technology being developed; however, UCD literature commonly recommends building and evaluating a minimum of three prototypes before deploying a technology into real-world applications [61]. The first prototype is used to present the concept of the device to potential users, to gather user feedback on initial ideas, and to generate potential alternative or additional functionalities and interface modalities. As it is intended for investigating a concept, first prototypes are usually put together quickly and without unnecessary complexity; these prototypes will typically be paper mock-ups and/or simple diagrams to demonstrate conceptual capabilities of the device, such as interfaces, data flow, and form factors. The second prototype incorporates what has been learned from the first evaluation to produce something that demonstrates more of the features of the proposed solution, but is not yet fully functional. Second-level prototypes commonly employ a "wizard-of-oz" approach [63], where the developer simulates aspects of the technology (e.g., the reaction of the device to a person making an error during a task) using button presses or simple scripts to demonstrate possible functionalities of the device. Evaluations with the second prototype will result in further details about the problem and the effectiveness of the new solution. With the knowledge gained and the investment of time and resources, subsequent prototypes have a level of detail and functions that are akin to what would be seen in the final product. The third and subsequent stages of prototypes are appropriate for in-depth, long-term clinical trials that are conducted in real-world situations with actual users to ensure the efficacy of the system and to help build a case for technology transfer and commercialisation.

3.2.3   EMPATHY-BASED DESIGN

Empathy-based design is a more recent human-computer interaction approach that stresses personal connection with users as part of the design process. More specifically, it emphasizes empathy, an "understanding of another person from the 'inside'", requiring an accurate sensing of a person, from the other's perspective, while maintaining one's own sense of self [64].

This approach is adopted to address design issues such as user diversity, user-developer differences, and emotion and experience design. Diversity applies to user groups such as older adults and those with disabilities who vary considerably in experiences, backgrounds, demographics, and health conditions [65]. Given such diversity, it can be difficult to find users who are representative, which is the cornerstone of UCD. Consequently, when designing for these groups, an empathic approach using the methods outlined below is recommended [66]. Empathy-based design helps to capture differences, where difference refers to situations in which developers differ significantly from users, such as designing for people from other cultures or for people with impairments. Finally, emotion and experience design is stressed when developing technologies or services that emphasize positive user experience over productivity, such as the case for social and game-based applications [67]. In all three of these situations, developing an empathic connection with people who are representative of end-users is the first step of the design process [68].

Two methodologies associated with empathy-based design have been applied to healthcare in particular. The first is user-sensitive inclusive design (USID), which is a methodology that emphasizes treating users as people first [66, 69]. Methods employed in USID include informal social gatherings to allow developers to get to know users on a personal level; and designing for *extraordi-*

nary users, which focuses on studying *outriders* in early design sessions [70]. Other methods include the use of theatre, which can range from short sketches to full productions with trained actors that conclude with interactive Q&A sessions with developers [71]. While theatre productions require planning and funding, they can be very helpful in exploring situations involving vulnerable groups in which stigma is an issue.

The other methodology is an empathy-based participatory approach [72], which has been applied to technology design for people with dementia. This methodology has its roots in participatory design, which requires significant user involvement throughout the design process and emphasizes a person-centered communication style. This style involves open, respectful, nonjudgmental communication with users [73]. Accordingly, participants' comments are taken at face value, with no need to support or defend themselves, even in the face of contrary evidence. By valuing the participants' emotions and experiences as they are expressed and without judgement, a developer can build trust and foster a supportive environment.

Regardless of the methods that are employed, open, accepting, genuine communication is the foundation of the empathic design approach. Empathetic-based design can elicit a deeper, more complex understanding of users, which can enable developers and designers to create technologies that better complement users' needs, abilities, and expectations.

Finally, it is important to remember that empathy-based design focuses on *understanding* an individual at a very personal level, requiring suitable approaches and techniques. UCD, in contrast, focuses on *analyzing* representative users by studying various aspects of those users (such as physical, sensory, motor, and cognitive aspects). The first emphasizes personal understanding of what an experience is like for a person; the second emphasizes analyzing various characteristics of representative users. Empathy-based design can be used in combination with inclusive or participatory design, as described above, or in combination with the understand-study-design-build-test cycle that is common to interaction design.

3.2.4   INCORPORATING PRIVACY IN THE DESIGN PROCESS

As touched on in our discussion regarding pervasive computing principles, it is vital to ensure appropriate security and privacy features are built into the design of any ZET as these technologies may gather and transmit potentially sensitive data about users. The application of sensors and related technologies to the provision of healthcare and rehabilitation brings additional considerations to the already complex issue of health information privacy [74]. Ensuring the privacy and security of user information will be a key determinant of the success of ZETs and other technologies.

In response to trends toward the autonomous capture and transmission of data, the Information and Privacy Commissioner of Ontario, Canada developed a guideline in the mid-1990s that outlines the concept of *Privacy by Design* (PbD). PbD entails embedding privacy into technologies beginning from the conceptualisation of the device, right through to commercialisation, retail, and ongoing operation, and support [74]. In conjunction with PbD is the notion of a *positive-sum paradigm*, whereby privacy, a user's wellbeing, and commercialisation of a device may all be supported if privacy safeguards are proactively built into a system from the outset [74]. Put another way, considering and addressing privacy safeguards appropriately from product conceptualisation onward is more likely to result in an acceptable, ethically responsible product and is generally easier than attempting to manage privacy post-hoc. This paradigm has been well received in the pervasive computing and ZET fields, and indeed, many researchers have made the argument for the necessity of creating positive-sum PbD technologies. For example, Coughlin et al. (2007) [75] found that the concerns of older adults with regard to smart home technologies included usability, reliability, privacy, and trust, among others. The study reported that pervasive systems would only be accepted if these issues were addressed right from the start. Research by Kotz et al. (2009) [76] described the goal of any remote health monitoring technology should be to develop usable devices that respect patient privacy, while also ensuring data quality and accessibility for the outcome of improved health. Corporations such as Microsoft and Apple have acknowledged the importance of PbD and claim to support its use in their products.

Six principles have been developed to help designers of ZETs and other technologies intrinsically incorporate privacy and security into a device throughout the design process and ensure that the positive-sum PbD paradigm is met. The purpose of the principles is to support and balance the needs of multiple stakeholders, who can include family, healthcare providers, technology manufacturers, insurance companies, government, in addition to the primary user themselves. More details about positive-sum PbD and the six principles outlined below can be found in [17].

Be Proactive, Not Reactive

The first PbD principle is to *be proactive not reactive* in the implementation of privacy features in any new technology by anticipating and preventing as many privacy issues as possible before they happen. The identification and prediction of possible privacy concerns users might have can be achieved through the employment of the various techniques described in the preceding sections, such as user interviews, focus groups, and other human factors techniques (e.g., role playing, simulations, etc.). Designers must also be aware of best practices regarding the secure handling of data and incorporate protocols that ensure the data that are captured, analysed, and transmitted by the system are handled appropriately. This includes not only adhering to current best practices, but to consider what is likely to happen in the near future so that technology can support retroactive upgrades, if possible.

Privacy as the Default

The second principle is *privacy as the default*, which seeks to deliver the maximum degree of privacy by ensuring that the default setting of the system is the one that ensures the maximum level of security. Put another way, a user should not be required to do anything to "turn on" privacy features in a system, rather the user should have to clearly and explicitly turn off any privacy features he or she feels are superfluous. Again, the technology designer needs to have a good understanding of the privacy and security features that should be in place by default, which settings can be deactivated by the user if they choose to do so, and what minimum level of security is required (regardless of any settings the user chooses) to ensure the data is handled in a way that is secure and compliant with relevant regulations.

Privacy Is Embedded into the Design

The third principle is that *privacy is embedded into the design*. This means that privacy features must not be add-ons or are included after the initial system is designed; they must be integrated into the architecture of the technology and the associated system support and business practices. The result of this principle is that privacy becomes an essential component of the core functionality delivered by the technology, without compromising operability.

End-to-end Lifecycle Protection

The fourth principle is *end-to-end lifecycle protection*. This principle advocates embedding privacy and security into the system prior to the first element of data being collected and to maintain privacy and security throughout the entire lifecycle of the data involved. This ensures that at the end of a system process (e.g., from data capture through the transfer of data from a home to a clinic, where after it is discarded) all data are handled and destroyed in a secure fashion. It also should consider how to handle data if the technology stops being used, malfunctions, is lost, or is stolen.

Visibility and Transparency

The fifth principle is *visibility and transparency*, which seeks to assure all stakeholders that a technology is operating according to the parameters stated by the technology's designers, researchers, and manufacturers. Furthermore, this principle includes a process for independent verification or audits to be made with respect to the security and privacy standards of the system. This principle is becoming an important aspect in the design and deployment of ZETs as these technologies are becoming more widespread and often handle sensitive data, therefore they are being subjected to increasing scrutiny from research ethics boards and governmental safety boards and associations (e.g., FDA, CSA) to ensure they employ appropriate security and privacy measures. At the same time, the zero-effort nature of their operation means ZETs tend to blend into the background, thus careful considerations must be made to ensure transparency about data and system security in a way that makes sense.

Respect for User Privacy

The final PbD principle is *respect for user privacy*. This principle describes the essence and motivation for PbD as, above all, PbD requires system designers and operators to keep the interests of the individual at the forefront of the design process. Respect for users' privacy can be accomplished by implementing measures such as strong privacy defaults, the ability for a user to learn about what measures are in place should they wish to do so, clear and appropriate notice of any changes to security or privacy settings, and the implementation of user-friendly options.

Applying PbD to ZETs

As per the UCD approach, the central concept in PbD is to focus on the user and his or her needs and preferences with respect to security and privacy issues. Privacy-friendly defaults, appropriate notice, and user-friendly options and interfaces in ZETs are important to ensure that the user can fully engage in the protection and control of his or her own personal information. Systems' functionality should be transparent and their components visible, particularly in the application of ZETs to home healthcare scenarios. For example, individuals who are being monitored by a ZET should be made aware of where the sensors are installed, what data is being collected, and who can access it. Moreover, users should be able to participate in this process, playing an active role in choosing what devices are installed and explicitly stating who should have access to resulting data.

Privacy and security options should be available to users (and any designated representatives) at all phases of their relationship with the device and the device provider, which includes prior to installation, during the initial evaluation of the technology, and while the device is operating on a day-to-day basis [74]. In line with the design approaches discussed in previous sections, engagement with users can and should begin at the initial design stages and continue throughout development. Identifying potential privacy concerns at a conceptual stage of design allows PbD to be applied as a technology is being built, creating the possibility for discussions regarding user requirements concurrently with design decisions. Moreover, incorporating privacy and security into each phase of the design process is generally easier and more effective than attempting to implement these practices after the technology has been built. Design practices used in the ZET field need to build complete visibility and transparency into systems by having representative users involved throughout the design process, by testing the final technologies, and by providing education sessions on how the system operates. This approach ensures that the system's users have an understanding of how a system operates and that reasonable and appropriate measures to protect privacy have in fact been incorporated into the final design [74]. Contrary to what one might expect, system transparency provides assurances that a person's privacy is being respected that foster greater technology receptivity. Moreover, affording the user a measure of control over his or her own data fosters trust in the device, promotes active participation in device use, and decreases the likelihood of device abandonment.

Once the user's privacy requirements have been determined, the next stage of applying the PbD principles is in the design of the ZET system itself. This requires privacy protections that are appropriate to the sensitivity and identifiably of the data and should be developed and proactively incorporated into the ZET. Given the (often) remote installation (e.g., a system running in a home rather than a clinical environment) and sometimes limited connectivity of the sensors themselves, it is important and far more effective in the long run to address all foreseeable privacy issues before they occur, as a breach in privacy would require updating or re-instrumenting a system after installation [74]. Wherever possible, protections should be embedded deeply into system components. For example, it may be possible to perform a great deal of data processing within the ZET itself or on a local (e.g., in-home) processor, which can drastically reduce the amount of data that is transmitted and allows the system to de-identify and encrypt data more effectively prior to transmission. By reducing the amount of information transmitted about the user, sensor, what is being measured, or any other kind of index that might connect the data to an individual, the odds of an unwanted third party intercepting and interpreting data decreases significantly. ZETs should also be designed to respect the principle of data minimisation, collecting only information that is required for the specified purpose. Data should not be collected, for instance, for unidentified potential future usage or because a developer or provider is interested in collecting data that are not related to the application in question. It is also crucial that systems are able to recognise if they are being compromised and alert the user and/or provider.

Finally, privacy features should require no or minimal effort of the part of the user; after settings are selected, the user should not have to actively enforce his or her privacy choices. As stated earlier in this section, ensuring adequate privacy and security is particularly important for ZET applications in the healthcare domain. Most ZETs are designed to operate without the user(s) being consciously aware of the systems, rather individuals are meant to be able to go about their normal lives, with devices taking readings inconspicuously and providing assistance only when required. Security and privacy measures for these types of systems are no exception, in that they should support the user continuously and discretely with little or no effort on the part of the user.

An interesting area of consideration is ZETs and other technologies that are designed to assist people in the case of an emergency. If a system is running or being used during an emergency, it is likely beneficial if the system can actively assist in procuring help. At this time, data privacy is likely not to be at the forefront of the individual's mind, although the data that could be used to assist the person could potentially be quite sensitive. For instance, a vision-based system could be able to automatically transmit images of the accident to emergency services so a professional can assess the situation and relay information to response crews, resulting in a faster, more targeted response. However, many people may see the automatic transmission of such information as a gross invasion of privacy. How these sorts of situations are treated, who gets to make ultimate decisions regarding potentially life-threatening situations, and if there are times when an individual's preferences should be overridden in the interests in protecting his or her wellbeing are some of the issues regarding privacy and security in ZETs and other smart technologies that are being intensely debated by the public, technology designers, and governments. As discussed in the following section about ethical considerations, these types of decisions can be more complex when people who may not be able to make informed decisions are involved, such as children or people with dementia. Regardless, privacy via embedded and by-default protections must be something that the individual can expect to be present and should be made aware of any exceptions, such as any differences in operation and data transfer during an emergency situation [74].

3.3   ETHICAL CONSIDERATIONS REGARDING ZETS

While technology can provide significant support, the ethical implications of their use must be considered from conceptualisation through design to implementation. Central to this is ensuring end-users understand what the technology does and that they are choosing to engage with it. This is especially important for ZETs as they can blend into environments and operate without conscious effort, which means people may be monitored or interact with them without being aware of it. Everyone who is a part of implementing a ZET (e.g., developers, providers, policy makers, and end-users) shares the responsibility of ensuring the ZET is ethically sound.

Ethics and technology is a complex topic. The form and function of technology is extremely diverse and rapidly evolving. The values and contexts of people and their societies are also broad and dynamic. While there can be no "one size fits all" approach to ethics and technology, general guidance exists to support its ethical development and use. One of the better-known and widely used ethical frameworks is Beauchamp and Childress' four principles of biomedical ethics [77].

1. **Respect for autonomy:** a norm of respecting and supporting autonomous decisions.

2. **Non-maleficence:** a norm of avoiding the causation of harm.

3. **Beneficence:** a group of norms pertaining to relieving, lessening, or preventing harm and providing benefits and balancing benefits against risks and costs.

4. **Justice:** a group of norms for fairly distributing benefits, risks, and costs.

To uphold these principles, people engaging with a ZET—from its designers to end-users —must understand how it operates and implications of its use to be able to estimate how the ZET relates to applicable contexts and values. This includes considering aspects such as who will use it, what it does, what type of data is being collected, and who has access to the data. A broad a range of stakeholders should be involved to enable a more holistic comprehension. Overcoming the challenges of discussing potentially complex issues with a diversity of perspectives results in superior products and systems; ones that are more acceptable to the end users and integrate more readily with regulatory requirements. From a developer's perspective, holistic comprehension supports ensuring the technology's design and intended operation are as ethically responsible as possible. From a policy maker perspective, it enables a better understanding of the technology landscape that enables more appropriate systems support for products and people. From an end-user perspective, it enables each person to more accurately weigh risks and benefits before freely choosing to engage with the technology, which is the essence of *informed consent*. Obtaining informed consent is widely considered an imperative of ethical technology use, however, how this is accomplished is highly dependent on the type of technology that is being used, its purpose, and the people using it. For example, a person living with dementia or a child will not be able to understand and consent to technology use in the same manner as a healthy adult. This does not excuse the need to inform and gain consent, rather it means that this must be done in a way that complements people's abilities.

As an in-depth discussion of ethics and technology is beyond the scope of this book, the reader is encouraged to explore the growing literature on the topic, such as Friedman and Khan's overview of values, ethics, and design in human-computer interaction [78], the characterisation of fairness in the assessment of technological design [79], and a proposed "ethical by design" manifesto [80].

3.4   KEY DESIGN CRITERIA FOR ZETS

The goal of a ZET is to enable the people using them to focus their efforts on achieving the intended task, not the operation of the technology. To do so requires designers to select hardware and software that appropriately complement the abilities of the technology users and contexts in which they are used. This can be a significant challenge as there is a plethora of approaches a designer could use to build each different system and sub-system of a ZET, as has been alluded to in previous sections. In this section, we present eight ZET design criteria developed by the authors based on the paradigms presented earlier in this book and the expertise of researchers in the field of smart technology development. These criteria are intended to guide technology designers who are working with special populations to ensure that the needs, preferences, and context of the user is taken into account. These principles are intended to be incorporated into any ZET and are applicable to all different aspects of a ZET, such as AI, pervasive computing, privacy by design, and, above all, sensitivity to user needs.

3.4.1   DEVELOP FOR REAL-WORLD CONTEXTS

ZETs should address a real-world need that is driven by the end user(s). Effective ZETs are not technology-driven; instead, they seek to understand and implement the most appropriate technologies for supporting the needs of the people who will use them.

Many assistive technologies and other devices developed for healthcare and rehabilitation are not adopted or are soon abandoned. It is estimated that anywhere from 50–70% of all devices are abandoned post-implementation [81, 82]. One of the most highly cited reasons for abandonment is that the technology did not meet the actual needs of the user; the device did not enable the user to do the things they expected the device would help them do and/or the device was more trouble than it was worth. To avoid such pitfalls, ZET designers need to first gain a good understanding the problem(s) for which they are developing the technology to solve. This includes the expectations of the end user(s), their capabilities, and the environment where the technology will be deployed. This understanding can be achieved through the use of the various techniques presented earlier in the book.

The first step in grounding the design process in real-world problems is to gain an understanding of the *context* for which the ZET is being designed. This includes how the technology fits into users' lives. The *International Classification of Functioning, Disability, and Health* framework published by the World Health Organization [2] defines contextual factors impacting uptake as external environmental factors (e.g., social attitudes, architectural characteristics, legal and social structures, climate, etc.), and internal personal factors (e.g., gender, age, disease type, coping styles, and education). The designer must also consider users' ability to acquire and implement the technology. For example, do the majority of target users have access to the Internet? Moreover, factors that are important to one user may be different from those that affect a different user even when their contexts seem very similar. As such, designers need to identify not only which factors are of importance, but the sensitivity of these factors; which are more robust/sensitive to variance across the target user groups. Applying this information to a ZET would allow the ZET to leverage commonalities and be sensitive to differences. Understanding real-world context is the basis for the designers to build the models, algorithms, and sensors for the system to be able to operate effectively.

3.4.2   COMPLEMENT EXISTING ABILITIES

ZET developers should focus on what the person is able to do and leverage these abilities to enable them to achieve what they cannot do.

The goal of ZETs, and indeed all technologies, is to enable people to do things they could not do otherwise. While providing support in a manner that supplants a person's abilities may achieve the same result as one that complements abilities, the emotional and physical outcomes may be drastically different. Research has shown that engagement in one's surroundings can have a significant impact on slowing the decline in abilities [83]. Technologies that support and complement existing abilities enable a person to remain as interactive in his or her environment as possible, fostering a sense of control and independence. For instance, many older adults with dementia have difficulties with mobility but are not permitted to use powered wheelchairs as impairments in judgement and reaction time can make them dangerous drivers. In response to this need, researchers are taking different approaches to assisted mobility that range from anti-collision systems to fully autonomous navigation. While the latter may be suitable for people who have lost all mobility, it is erroneous to assume that older adults with dementia do not have the capacity to know where they want to go or how to operate their chair to get there. With a fully or mostly automated approach, the chair's occupant becomes little more than a passive passenger, placidly being conveyed to destinations with little control as to where they are going or how they get there. This approach decreases cognitive stimulation and sense of mastery or control, which increases feelings of helplessness and apathy. On the other hand, intelligent chairs can complement the driver's abilities to promote feelings of dignity and independence, such as a chair that provides prompts to help its driver make navigation decisions and the use of anti-collision devices to avoid hitting people or objects. This approach enables a person to assume control and ownership over where they wish to go without the risk of hitting people or objects.

3.4.3   USE APPROPRIATE AND INTUITIVE INTERFACES

Interfaces should be simple, intuitive, and appropriate to make the ZET more accessible and useful to the people who use them. Whenever possible these interfaces should blend into the user's existing environment and context.

To be zero-effort, the ZET must complement the users' abilities, including physical, cognitive, and sensory abilities. Technologies designed for users with special needs should take into account the particular features that would be most appropriate and useful. If a user interface is required, a large display with a simple, high contrast interface is an example of design features that would support older adults. While touch-screen mobile phones may provide a convenient and adaptable interface, they are not always appropriate as they may be unfamiliar to some users (e.g., older adults, although this trend is changing), can be easily misplaced (e.g., someone with memory impairment), and are potentially difficult or finicky to use (e.g., people who have poor eyesight or a tremor). Ambient-embedded technologies, where the technology is integrated into the environment and operates without an explicit interface, is an alternative that is growing in popularity as they do not require the user to remember to wear a device; the device engages with the occupant as they go about their everyday lives. Moreover, embedded technologies maximise connectivity while minimising maintenance, such as changing batteries or accidentally damaging the device. Reacting to trends in the growing role of ambient monitoring in healthcare, most assisted-living communities are being built to support embedded sensor technology with extra power, communication conduits, and removable paneling to install and reconfigure embedded technologies more easily. In general, while ambient technologies are gaining in popularity, they have their drawbacks. For example, the technology may be fixed to monitoring the area where it is installed, and monitoring multiple occupants still poses challenges for many systems. At the end of the day, there is no one-size-fits-all approach to creating a ZET interface. Rather, the interface should be designed to complement the abilities of the person and provide support for their tasks with effortless operation of the device.

3.4.4 ENCOURAGE USERS' INTERACTION WITH THEIR ENVIRONMENT

*Devices that encourage people with disabilities to interact with their environments can stimulate interest in their surroundings and help them to relate with other*s.

Changes in motor skills, sensory perception, or cognition can result in a profound loss in control over one's environment. This loss of control can lead to diminished or impaired physical and social involvement. As will be illustrated in the case studies later in this book, many ZETs are targeted toward ameliorating these losses through interventions that range from supporting activities of daily living to promoting social inclusion. For example, the COOK project has been designed to provide aid for people with traumatic brain injury in cooking a meal autonomously [84, 85]. The device consists of a touch screen interface that allows users to select what type of meal they wish to cook and offers supportive tips and reminders to individuals throughout the cooking process. If a potential risk of danger is detected (e.g., unattended food cooking on stove) and cannot be resolved despite requests to the user, the safety system can automatically turn off the stove and contact assistance as necessary. This interface has been designed to be friendly and easy to use by people with traumatic brain injury and in future the researchers are considering modifying this to be used by older adults with dementia. Being able to cook a meal independently is a significant activity that would support an individual's sense of autonomy. Louie, Vaquero, Nejat, and Beck (2014) [86] describe how they have developed algorithms to enable a Bingo facilitating robot, Tangy, to schedule a Bingo game with seniors; remind the participants that the game is about to start and the location where the game will take place; and facilitate the Bingo game while providing assistance with multiple players in real time. These algorithms help determine what would constitute appropriate behaviour for the robot when interacting with seniors, as well as how best to facilitate the Bingo game and provide assistance. These are examples of ZETs that help to complement the users' abilities by encouraging them to interact with the technology and their environment through usable and useful interfaces.

3.4.5   SUPPORT CAREGIVERS

Ensure the technology supports the caregiver in a way that reduces caregiver burden and enables them to focus on other tasks, such as building a better relationship with the person they care for.

Caregivers (i.e., someone who provides care to someone else) are often considered primary or secondary users of a ZET, thus the technology should operate in an intuitive fashion for them. Caring for someone with a disability can be a challenging task that can result in a great deal of physical and emotional demands, which have been shown to cause high levels of stress, depression, and even an increased risk of mortality [87, 88]. A device that adds burden is unlikely to be accepted as the caregiver is often already at the limits of his or her capabilities. Moreover, the caregiver may have morbidities of their own to manage. As pre-existing demands are often already overwhelming, requiring a caregiver to learn or work a complex technology is an impractical alternative. Thus, the ZET should be as robust and as autonomous as possible, requiring only simple and infrequent input from a caregiver in a way that reflects their wants and needs. Ensuring the caregiver understands how a technology works, is supported by the technology, and does not need to invest effort in a technology will foster confidence and acceptance of the intervention.

3.4.6   COMPLEMENT EACH INDIVIDUAL'S CAPABILITIES AND NEEDS

Leverage artificial intelligence, machine learning, and decision making techniques to create technologies that are able to complement users' specific and individual abilities, enabling guidance that is appropriate, relevant, emotionally aligned, and effective.

Like any other person, users of ZETs have their own wishes and preferences about the type of technology they would like to use, what it looks like, how they would like to interact with it, and how they would like to feel while doing so. Identifying these preferences is important, particularly for technologies that actively interact with the user (as opposed to passive monitoring). People with various disabilities, especially cognitive impairments, are often unable to reliably answer questions or perform specific actions, therefore, capturing and understanding preferences can be challenging as they are often inferred from observations over a period of time. In this respect, work in the field of preference elicitation is very valuable (as discussed earlier in this book). An example of this is the autonomous identification of preferences though computer-based observations of a person's actions. Knowing person's preferences would help to enable device personalisation and minimal explicit demands from the caregiver and care recipient. It is through the identification and accommodation of user's individual abilities and supporting their needs (emotional as well as physical) that a technology becomes truly useful.

3.4.7   PROTECT USERS' PRIVACY AND ENABLE CONTROL OVER PREFERENCES

Devices should operate in a way that matches the user's information sharing preferences, needs, and abilities. The user (or his or her caregiver) should be able to easily change these settings at any time to match changes in circumstances.

Technology should be transparent so that it allows stakeholders (e.g., the primary user, his or her caregiver, family members, health practitioner, etc.) the ability to understand and have control over the data that are being collected, stored, and transmitted. A high-level understanding of what is going on inside the "black box" makes people feel more comfortable with the technology and therefore more likely to accept it. Moreover, implicit clarity and transparency about what the device is doing enables people to intuitively gauge the value of the assistance that is being provided to them and affords them the ability to adjust it to fit their needs, making it more applicable over the long-term. There is also a wide variability in peoples' comfort level toward what data is communicated. As such, users should have control over settings so that they can select the configuration of their choice.

3.4.8   ENSURE EXPANDABILITY AND COMPATIBILITY

ZETs should be compatible with other technologies and, whenever possible, a ZET should be easily expandable to include other required functions.

As dictated by the principles of pervasive computing, ZETs should be able to operate as stand-alone devices or in tandem with other technologies. This capability allows consumers to pick which ZET(s) are appropriate for their particular needs and addresses practical challenges associated with developing a single technology that can support multiple daily activities. The ability to combine data from multiple technologies also allows for data fusion and a richer, more holistic representation of the people they are supporting. This information can be used to provide targeted and tailored interventions to match the individual's needs and particular situations.

CHAPTER 4

Building and Evaluating ZETs

Guided by the methods described in the previous sections, device development is usually iterative in nature, becoming more involved as a product evolves. Development is often done in a modular fashion, with a number of sub-systems interacting to form the overall device. Each sub-system is usually responsible for a specific aspect of the device, such as sensor input or decision making, and communicates relevant data to the other sub-systems over a shared bus or network. This approach is flexible and allows designers to develop, integrate, and optimise sub-systems sequentially or in parallel without having to retool the entire device. It also allows sub-systems to be altered, adapted, and upgraded to fit specific applications or to incorporate new algorithms or hardware as they become available.

Regardless of the approaches used to create or improve a ZET, developers should set measurable outcomes; specific goals or objectives that are used to gauge the extent of development and implementation efforts. As outcomes represent goals the design team wishes to achieve, they can be used to guide the direction of a technology's development, gauge performance improvements, and signal when milestones have been reached. Outcomes can be short-term or long-term, can be ranked in terms of importance, and may change over the course of the development process as a project's scope becomes more defined.

Importantly, outcomes should include non-technical measures of success, such as the device's impact on user independence and/or quality of life as well as user satisfaction with the device. While these outcomes are subjective and therefore difficult to measure quantitatively, they are crucial as they signify the ultimate measures of a device's success. For instance, if a device operates flawlessly from a technical perspective, but users do not feel it is useful, or worse, if it causes users to become annoyed or frustrated, then it is not a successful intervention; it introduces challenges rather than solving them and will likely be rejected by the people it is intended to help.

The techniques listed below are not an exhaustive list, rather they are intended to give examples of different approaches. Depending on the team's resources, stage of development, and nature of the application, device developers may use some or all techniques sequentially, in parallel, or iteratively throughout the design process.

4.1 IN SILICO TESTING

In silico testing consists of creating, debugging, and optimising a new technology in a virtual environment on a computer. The purpose of in silico testing is to create an operational model of the system as well as to identify and address as many issues as possible in a virtual environment before porting the system to a version that interacts with hardware. A system solely in a virtual environment has the advantage of being extremely malleable and allows developers to modify the system as they wish without the added complications, such as ensuring changes do not interface with hardware or interfaces. Another advantage to in silico testing is that there are no physical constraints to the system. Using this technique, different approaches can be investigated relatively easily and at a low cost prior to building a physical prototype. Indeed, the final virtual model may dictate hardware and interface choices in unforeseen ways. For instance, while a developer may have a particular application and associated hardware in mind, such as the type of sensor to be used, the development and testing of a system in silico can help to solidify specific physical deployment parameters or may point to an entirely different sensor or approach than the one that was originally envisioned.

As in silico testing uses explicit and known inputs, it is a powerful and effective method for adding new functionalities, targeting areas for development, debugging and optimisation, and investigating specific scenarios. Developers can run an identical input as many times as they wish, enabling the direct comparison of different versions or approaches to solving the same problem. For example, in computer vision applications, developers commonly use machine learning with large datasets of pre-recorded videos or images to train a system. Using the same dataset allows developers to try different machine learning techniques and to tune the application's parameters to optimise factors such as accuracy and computational efficiency, and balance trade-offs, such as model complexity and precision versus generalizability and resources required. As described in the section on AI earlier in this book, cross validation is a technique that is often employed, where a portion of a dataset is set aside to validate the model after it has been trained. With cross validation, the model is trained using most of the data set then tested on the reserved data to gauge the model's performance and applicability to the application.

Once a system model has been selected, developers often use button presses (e.g., using a keyboard or buttons), scripted code, or a sub-set of a dataset (as discussed above) to simulate input the device could experience if it were operating in the real world. Sometimes experts in a related field, such as clinicians, will be consulted to construct realistic test scenarios. This final step of in silico development can be used to further optimise a system before it is ported to a prototype.

4.2   BENCHTOP TRIALS

In benchtop trials, hardware and software are combined to form a functional version of a device that developers can interact with. The ability to interact with a device and its components adds a layer of complexity as the device's hardware and software are physically present, integrated, and, ideally, running in real time. Benchtop trials allow developers to physically simulate scenarios in a manner that is representative of what the device would experience in a real-world application. Interacting with the device assists with the identification and tuning of numerous design requirements, such as the hardware selection, logistics concerning device assembly and installation, and the establishment of communication protocols between device components, sub-systems, and external services. Devices can be assembled piecemeal, with one portion or sub-system of a device assembled at a time. Often, the portions of a system that are not yet built can be simulated in silico and interact with the parts that are physically present.

Benchtop trials can range from testing the system and its components at a workstation, to testing in mock-ups of real environments. Many organisations are using test smart homes or apartments to trial new devices in a simulated home environment [89, 90], which can have researchers or participants from populations of interest simulate the activities the ZET will support. This enables the prototype technology to be evaluated on its own or with other systems, both smart and otherwise. Through benchtop trials, a device transitions from the conceptualisation phase into an operational device.

4.3   ACTOR SIMULATIONS

While experienced developers may have a good understanding of the general behaviours and attitudes of the population they are designing for, these traits can be quite difficult to replicate authentically. Actors, however, have trained extensively to emulate emotions and attitudes to play specific roles. Role-playing by actors is often used to simulate conditions or situations of interest for teaching and assessments in medicine. This allows medical students to learn about specific scenarios and investigate the effects of treatment options prior to encountering them in a real patient [91]. Similarly, using actors has been used to optimise devices and assistive technologies. For example, in the area of "affected computing," actors have been used to better simulate expressive or emotional speech [92, 93]. The use of actors can also allow developers to gain a better understanding of how a device is likely to be used and how it will react to the population of interest prior to deployment.

Using actors is a valuable and cost-effective option compared to testing solely through real-world trials (which are discussed below). Once actors are trained on examples of the population of interest, they can simulate scenarios consistently for as long as is needed. This is especially advantageous when the targeted user group is a vulnerable or frail population, such as older adults with dementia. Using actors allows data to be collected much more quickly, has a low ethical risk (as actors are cognitively aware, consenting adults operating in simulated situations), generally requires significantly less resources to complete, and allows for the emulation of specific scenarios, including those which may be rare but significant in the real world. The ability for actors to replicate specific behaviours and scenarios gives developers the benefit of being able to test a device or system with more conditions and more replications than might be encountered using a small clinical trial or pilot study.

One study has examined the use of actors to develop ZETs for older adults with dementia by having professional, older adult actors train on videos of people with dementia washing their hands [94]. The actors then emulated handwashing while first being guided by a human caregiver, then being guided by COACH; a ZET that assists people with dementia through handwashing (as described in the case studies later in this book). Professional caregivers were asked to watch videos of the actors and (previously captured) videos of older adults with dementia then asked to rate how believably the person in the video portrayed dementia. Not only did the professional caregivers give similar believability ratings for both conditions, the actors elicited responses from COACH that were comparable to real older adults with dementia [94].

While using actors is a powerful method of prototype device debugging and optimisation, this method does not replace the need to test the device with real people from the target population(s). Rather, optimisation using actors can improve device performance compared to benchtop trials alone, resulting in better performance in real-world trials and in the ultimate deployed system or device.

4.4   REAL-WORLD TRIALS

Trials with a representative group of users from the population(s) of interest are implemented in the later stages of the development process as they involve the device being tested with people and in the environments that they are designed to operate with. No matter how thoroughly a device is scrutinised prior to clinical trials, the technology will likely be confronted with a multitude of scenarios that have not yet been encountered. As such, it is only through the exposure to a representative sample of end-users and environments that developers can truly gauge how the technology will perform in the real world.

Real-world trials range from pilot studies in controlled environments to large-scale, long-term installations in peoples' homes. Conducting real-world trials is the most direct and comprehensive method of testing a device's capabilities and robustness, however, they are also usually far less supervised than other forms of device testing, therefore the device must be robust and functional prior to deployment.

While real-world trials are a crucial step in the development process, they are also the most complex and costly. A technology that is going to be used in a real-world trial should be optimised as much as is reasonably possible prior to the trials to maximise the likelihood of the technology's success. Methodologies should be carefully planned to ensure that trails run smoothly, that there are sufficient resources available to complete the trials, and that data is captured that allows for a critical and meaningful evaluation of the device's performance. It is important that real-world trials are well supported to ensure logistics are handled in a timely and professional manner, including equipment installation and removal, user education, and support in the case of device failure. Not doing so may negatively bias participants quite significantly, which may cause the intervention to be poorly rated when it may have succeeded otherwise. Moreover, contingency support must be in place to ensure that no harm comes to trial participants should the technology fail; this is especially important if the device is intended to support a person's wellbeing. In particular, trials involving people with disabilities generally require special considerations and extra resources. These demands can often result in a research team that is larger and from a greater diversity of backgrounds than the device development team.

By and large, the challenges presented by real-world trials are outweighed by the opportunity for developers to gain a good understanding of how the device operates when it is deployed with representative end users interacting with the technology in real environments. The results and feedback gained through real-world trials can be especially helpful in identifying deficiencies, are very helpful in targeting future development efforts, and can provide the proof of efficacy required to transfer a device from the research and development phase to commercialisation.

CHAPTER 5

Examples of ZETs

Over the past several years there have been numerous technologies and systems developed for people with a wide variety of disabilities. As such, this section will only include examples of technologies that meet the previously stated definition of a ZET, specifically *technologies that employ techniques such as AI and unobtrusive sensors to autonomously collect, analyse, and apply data about the user and their context*. In addition, most of these systems have been developed for people with cognitive impairments because this population will most likely benefit the most of these types of technologies and the adaptability, guidance, and compensatory abilities they may provide. This section reflects the current trends in ZET development and focuses on devices used to support people with various rehabilitation needs. The reader is referred to [95–98] for more information on intelligent technologies and systems that fall outside the scope of this book.

5.1   AREAS OF APPLICATION

The complexity and variability of a disability means that support must be appropriately sensitive, personalised, dynamic, robust, and context-aware. This can be especially true for people with a cognitive impairment, as users can lack the understanding or judgement required to identify and rectify a situation if the ZET provides sub-optimal guidance or, worse, makes a mistake. However, the benefit of using ZETs can also extend to other populations, including the general public. The following six areas have been identified where ZETs can provide support.

1. **Safety:** A cognitive impairment can often compromise a person's safety as he/she may not be able to anticipate or react to adverse situations appropriately. Monitoring a person can enable the detection of adverse events, such as the person falling or becoming ill, and procure assistance when required.

2. **Long-term trend prediction:** Monitoring a person over an extended period of time can be used to detect long term trends in behaviour. Deviations in trends can indicate potential health problems or changes in cognition.

3. **Assistance with activities of daily living:** Being able to perform activities of daily living, such as bathing, toileting, and dressing are crucial to maintaining independence. Difficulties with executive memory functioning often results in an inability to remember what steps are required to complete an activity or how to complete them. Supporting daily care activities can help to mitigate dependency on a caregiver.

4. **Communication:** Cognitive impairment can make interpersonal interactions difficult, causing feelings of isolation and depression. Technologies that assist in recollection and promote communication can enable meaningful interactions with people who have a disability.

5. **Leisure:** Participating in leisure activities is an important way for people to relax while simultaneously encouraging personal expression. Participation in leisure activities, such as activities of artistic expression, can provide a meaningful occupation for people with a cognitive impairment, increase their engagement with their environments, and provide an emotional outlet.

6. **Cognitive stimulation:** There is growing evidence that cognitive stimulation can help to slow the progression of impairment. Technologies that foster engagement could play a significant role in slowing cognitive decline and in the monitoring of cognitive abilities.

It should be noted that these areas are not mutually exclusive. As will be shown in the following examples of different technologies, one type of ZET can span more than one of these application areas.

5.2   OVERVIEW AND COMPARISON OF EXAMPLES

Table 5.1 summarizes the systems described in this section, with examples that include a variety of application areas, pervasive computing devices, and AI methodologies. Regardless of the technical approach, all of these systems are intended to support users in their goals to interact with their environments; none are meant to provide fully automated, context-free solutions. As illustrated by Table 5.1 and the subsequent discussions of each example, the majority of research in this field is on technology for adults and has especially focused on older adults with cognitive impairments, such as mild cognitive impairment or dementia. In terms of testing, each of the technologies described in this section are at differing stages of design and development. Some technologies have undergone technical efficacy trials in the lab, others have undergone trials with targeted end-users in-situ, and a few have reached the consumer market.

Table 5.1: A summary of the ten ZET examples described in this section

Features / Projects	Application Area	Target Population	Design Paradigm	Input Sources Devices	Output Devices	Artificial Intelligence and Computational Methods	Development Stage
DAAD (Section 5.3)	Safety (Detection and Prediction of Agitation and Aggressive Behaviours)	Adults (advanced dementia)	User-Centred Design	Multi-modal: Embedded Sensors (camera, accelerometer (wearable) door/motion, pressure mat)	Action output (Data sent to care provider computer)	Feature extraction to develop ML classifiers (DL) and generalizable predictive models, using a discriminative and (in future) generative abstraction approach	Concept Development & Usability Study (In-institution trial with people living with dementia)
				Project Age: 2 years			
				URL: www.researchgate.net/project/Agitation-and-Aggression-Detection-in-Older-Adults-with-Dementia			
Ambient Vital Signs Monitoring (Section 5.4)	Long-term Prediction (Vital Signs Monitoring)	Adults (heart failure)	User-Centred Design	Embedded Sensors (measures heart rate and other characteristics, body weight and temperature, blood pressure)	Action output (Data sent to care provider computer) Using discriminative abstraction approach currently.	Feature extraction and comparison of data to gold standard signals.	Concept Development & In-lab Pilot (Simulated user trials with individual device prototypes with young adults)
				Project Age: 8 years			
				URL: www.iatsl.org/projects/archived/heart_failure.html			

Non-Contact Visual Cardio-pulmonary Monitoring (Section 5.5)	Long-term Prediction (Sleep Apnea)	Neonates to Adults (who may have sleep apnea)	Universal Design	Environmental Sensors (Vision-Based IR Camera)	Action Output (Data sent to computer)	Computer vision techniques using feature points on video images, principal or independent component analyses, and spectral analysis. May use (in future) supervised discriminative abstraction approach.

Concept Development & In-lab Pilot (In-lab trials with adults in five simulated sleeping positions)

Project Age: 5 years
URL: www.iatsl.org/projects/sleep_apnea.html

COACH (Section 5.6)	ADL Support (Hand Washing)	Adults (cognitive impairment, dementia)	User-Centred Design	Embedded Sensors (Vision-Based Web Camera, microphone)	Audio, visual, and action outputs (Speakers and flat screen monitor, data to computer)	Decision making using a POMDP and computer vision (3D skeleton and object tracking). Adaptive to user preferences, culture and emotion, and is event based.

Concept Development & Usability Studies (In-institution trials with adults with moderate to severe dementia)

Project Age: 20 years
URL: www.iatsl.org/projects/intell_env.htm

Culinary Assistant and Environmental Safety System (Section 5.7)	ADL Support Safety Communication (Meal Preparation)	Adults (cognitive impairment, traumatic brain injury)	User-Centred Design	Embedded Sensors (Intelligent Stove, Doors, Windows)	Audio, visual, and action outputs (Speakers, lights, output screen, appliances, data sent to care provider computer, mobile phone)	Computer vision techniques including motion analysis, object/activity recognition. Uses hierarchical decision tree state spaces.

Concept Development & Usability Studies (In-institution trials with adults with cognitive impairment)

Project Age: 5 years
URL: www.usherbrooke.ca/domus/fr/recherche/nos-projets-de-recherche/cook-assistant-culinaire/

PEAT (Section 5.8)	ADL Support (Reminder System)	Adults (cognitive impairment, executive function impairment)	Scenario-Based Design, User-Centred Design	Embedded Sensors (Wearable RFID, motion, contact, GPS sensors)	Audio, visual, and action output (Light/sound, visual display and data to cell phone)	Planning and activity recognition algorithms. Uses unsupervised-interleaved Hidden Markov Model, Propel AI, Option-based hierarchical POMDP, and timing.	Product in Market, Product Development, & Usability Study (In-institution trials with patients with cognitive impairment)

Project Age: 21 years (since first version launch)
URL: brainaid.com

BRAZE™ Obstacle Detection Systems (Section 5.9)	ADL Support Safety (Obstacle Avoidance)	Children to Adults (who use a wheelchair)	User-Centred Design	Embedded Sensors (ultrasonic)	Audio, visual, and physical output (Lights, beeping, vibration)	Context awareness by use of location and distance sensors.	Product in Market, Product Development & Usability Study (In home and institution trials with people with mobility difficulties)

Project Age: 2 years
URL: brazemobility.com

HELPER (Section 15.10)	Communication Safety (Fall Detection)	Adults (living at home, healthcare or long-term care facilities)	User-Centred Design	Embedded Sensors (Camera, microphone)	Audio output and action (Sent to speakers, message sent to phone, email, or smart phone app)	Machine learning via random forests and computer vision. Natural language processing and spoken dialogue system recognizes "yes" and "no" user responses.	Concept Development & Usability Study (In-home and institution trial with healthy adults and older adults with cognitive impairments)

Project Age: 12 years
URL: www.iatsl.org/projects/ers.htm & www.iatsl.org/projects/speech_recognition.html

Ambient Activity Technologies (Section 5.11)	Leisure & Cognitive Stimulation (Music, games pictures, objects for touch)	Adults (cognitive impairment, dementia)	User-Centred Design	Embedded Sensors (Portable RFID)	Audio and visual output (Speakers, lights, and tablet screen)	Aware of different users and adaptive to user via preference elicitation (history tracking).	Product in Market (ABBY), Product Development, Concept Development (Centivizer) & Usability Studies (In-institution trials with long-term care home residents)

Project Age: 3 years
URL: Abby – www.ambientactivity.com; Centivizer – www.centivizer.com

Autonomous Assistive Robots (Section 5.12)	ADL Support (Meal Preparation); Leisure & Cognitive Stimulation (Bingo Game and Trivia)	Adults (cognitive impairment, dementia)	User-Centred Design and Empathy-Based Design	Embedded and 3D Sensors (camera, microphone, tablet, laser range finder)	Audio and visual output and action (Speakers, tablet screen, robot actions including gestures, body language, and facial expressions)	Uses planning and scheduling algorithms, finite state machine techniques, ML, facial features and gesture recognition and affective computing, object recognition, natural language processing.	Concept Development & Usability Studies (In-institution trials with long-term care home residents, staff, and family members)

Project Age: 13 years
URL: http://asblab.mie.utoronto.ca/research

5.3   THE DETECT AGITATION AGGRESSION DEMENTIA (DAAD) MULTI-MODAL SENSING NETWORK

Contributor: Dr. Shehroz Khan, Intelligent Assistive Technologies and Systems Lab, Department of Occupational Science and Occupational Therapy, University of Toronto

This example of a ZET falls under the application area of "safety". Dementia is a neurogenerative brain disease that affects nearly 50 million people world-wide with nearly eight million new cases identified per year [99]. In Canada, according to the Canadian Institute for Health Information (2010) [100] more than 50% of individuals living in long-term care have dementia. At some point in the disease, many people living with dementia will experience episodes of agitation and aggression [101, 102]. In institutionalized settings, these feelings of agitation and aggression may lead to response behaviours (e.g., screaming, hitting) which in-turn may result in negative confrontations between residents and staff, sometimes resulting in injury or death [103–105]. The detect agitation aggression dementia (DAAD) platform is a multi-modal sensing network of devices developed to better detect and predict when a person living with dementia may be feeling agitated or aggressive [106]. The DAAD platform uses a collection of embedded sensors including cameras, wearable devices, pressure mats, and motion and door sensors to monitor the activity, location, and physical condition of people living with dementia. The data from these sensors is used collectively to automatically detect and predict incidences of agitation and aggression in persons living with dementia in a long-term care facility [106]. Physical movement (e.g., accelerometer data), physiological changes (e.g., heart rate) activity (e.g., bathroom door opening), and sleep quality (e.g., changes in position, number of bed exits, respiration, heart rate) are measured and paired with location information (e.g., from camera video) and clinical notes to identify incidences of agitation and aggression.

A wrist device worn by patients has accelerometer(s) and sensors to capture motion-based activity and blood volume pulse (see Figure 5.1a). A pressure mat is placed under the patient's regular bed mattress (see Figure 5.1b), motion and door sensors are placed in the bathroom room (see Figures 5.2a and b), and cameras are installed in the institution hallways where the patient's reside (i.e., in public spaces) [106].

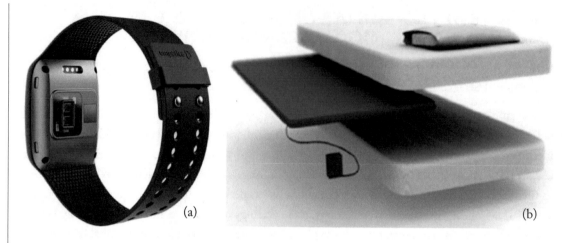

Figure 5.1: (a) On left, wrist device worn by patient (photo credit: https://www.empatica.com/research/e4/); and (b) on right, pressure mat (photo credit: http://quantifiedself.com/2011/11/toolmaker-talk-rich-rifredi-bam-labs-2/).

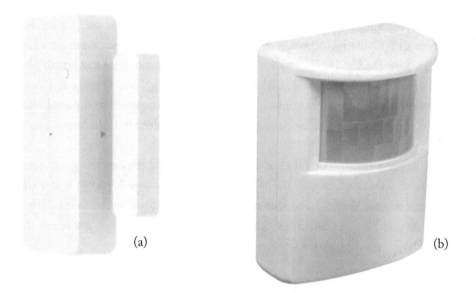

Figure 5.2: (a) Door sensors (photo credit: http://www.insteon.com/open-close-sensor/); and (b) Bathroom motion sensor (photo credit: www.insteon.com/motion-sensor/).

Currently, the DAAD is still in the early development phase and sensor data are transferred manually or sent wirelessly via email alerts to a computer for processing and analysis. This processing involves extracting features from raw sensor data, which could indicate the presence of agitation or aggression. These features will be used with DL methods such as recurrent neural networks that can model the temporal nature of the data,to build machine learning classifiers and generalizable predictive models. These classifiers and models will then be used to automatically identify similar patterns in future data [106]. In the future, discriminative or generative abstraction approaches may be used to handle a larger volume of output data. The sensor data would also be transferred automatically to a computer for processing, analysis, and interpretation. Since late 2017, data has been collected from patients living with dementia in a pilot study from a Geriatric Psychiatry ward of a large urban rehab hospital. See Khan et al. (2017) [106] for further details. The component devices have been successfully tested for function on adult users in a private home.

Following the design principles used in the development of ZETs, the devices used in the DAAD have been carefully selected for use by the intended target group after considering their habits and abilities in a real-world context. The individuals being monitored with this system will not need to alter their behaviour in anyway in-order for this system to be used and to operate.

The next two examples illustrate the use of ZETs in the application area of long-term prediction.

5.4   AMBIENT VITAL SIGNS (AVS) MONITORING

Contributor: Isaac Sung Jae Chang, Intelligent Assistive Technology and Systems Laboratory, Department of Occupational Science and Occupational Therapy, University of Toronto.

The ambient vital signs (AVS) monitoring system, is comprised of a collection of ZETs that work together to unobtrusively monitor a person's vital signs. Considering cardiovascular health, conventional monitoring devices for cardiovascular disease typically require active physical engagement, demands consistency, cognitive capability, and a willingness on the part of the user to properly operate these devices. This cognitive burden can be exacerbated in people with cognitive impairment as studies have observed these issues in cognitively healthy individuals [107]. For example, in order to monitor one's blood pressure, the person must know how to use the blood pressure machine properly and remember and be willing to use it. In the case of wearable devices such as a watch that can measure heart rate, it must be worn on the wrist to collect signals. If the person being monitored forgets to wear the watch, it cannot properly monitor the heart rate. One solution to these issues of user non-compliance, stigmatisation, and incorrect usage, is to collect vital signs unobtrusively, using ZETs, while the person being monitored performs his/her activities of daily living (ADL) [108], which is what the AVS system will do. The AVS system uses sensors embedded in objects commonly found and interacted with in a home to collect vital sign information from the person being monitored. Currently, the AVS monitoring system consists of five devices used for measuring heart rate, electrical signals from the heart (i.e., using electrocardiogram—ECG), approximate blood pressure, body weight, mechanical characteristics of the heart (i.e., using a ballistocardiogram

(BCG), and body surface temperature (i.e., using a thermal camera). These devices have sensors embedded into objects such as: a floor tile, chair, bed, and blanket. A summary of the parameters collected by each home object can be found in Table 5.2.

Table 5.2: Ambient vital sign monitoring system and the parameters monitored by each device

Device	Parameters Monitored
Floor tile	Heart rate, electrocardiogram, mechanical characteristics of the heart, approximate blood pressure
Chair	Heart rate, electrocardiogram, mechanical characteristics of the heart, approximate blood pressure
Bed	Weight, heart rate, mechanical characteristics of the heart
Blanket	Respiration
Thermal camera	Surface body temperature

As illustrated in Figure 5.3, the goal of the AVS monitoring system is to observe the short and long-term trends in collected vital signs and to notify caregivers or clinicians if signs of deteriorating health are observed. As the person being monitored interacts with these objects while performing ADL, the devices collect vital signs to assess his or her health. By doing so, the burden of physical and cognitive engagement of the person being monitored to operate conventional monitoring devices is reduced to zero or close to zero.

Although this system is still at an early stage of design and development, individual device prototypes have been tested with healthy adults and older adults with heart failure (HF). This testing is important to validate and characterize the ability to collect vital signs in this manner. Field trials with healthy adults demonstrated that a person's heart rate could be tracked and ECG signals measured through a floor tile, all with fairly high accuracy.

Trials with older adult participants with HF are also being conducted. Preliminary results show that their weights can be measured through a bed and heart rate can be measured while seated on a chair with very low error rates. As such, the AVS monitoring system shows potential in monitoring vital signs with zero-effort.

For the studies performed, the participants were not asked to use any type of conventional vital signs monitoring equipment except for gold standard equipment and the parameters were entirely measured with zero-effort, meaning the person did not need to learn to use any devices or how to use the AVS system. The perception of the participants on the zero-effort devices were also studied where the participants (i.e., older adults with HF) expressed positive opinions about the system such as increased sense of security, reassurance, as well as unobtrusiveness [109]. Some concerns were raised as well such as affordability, privacy, multi-user, and user preference issues. These are also pointed to by the design criteria of ZETs and should serve as guiding points in future developments. For example, while the monitoring system is unobtrusive and may not be seen, the person being monitored should be fully aware of what is being installed, what will be measured, and who will have access to the data, and have control over the entire process.

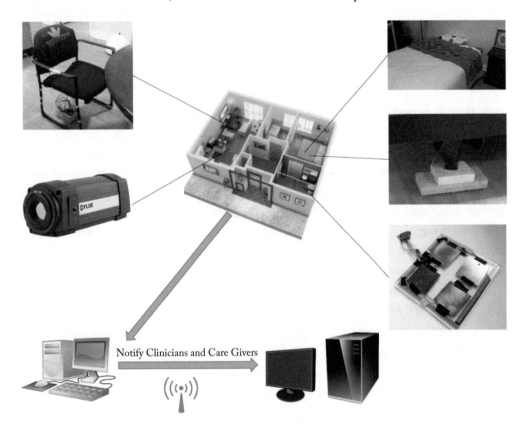

Figure 5.3: Ambient vital sign monitoring system. The system will collect vital signs and notify clinicians and/or caregivers if abnormal signs are observed (photo credit: Intelligent Assistive Technology and Systems Laboratory, University of Toronto).

In the current state, the technology cannot differentiate between vital sign changes resulting from an adverse versus non-adverse health event such as an increased heart rate due to a health abnormality or due to regular exercise. To address this issue, the system will be paired with an activity monitoring device or incorporate complementary sensors (e.g., computer vision or radar) to increase knowledge about what the person is doing and thus distinguish between different contexts. Future development on this system will also include using a discriminative abstraction approach to manage the data and investigate various ML techniques that may be suitable for later stages of technology development. Similar to the previous ZET technology, the development of the AVS monitoring system follows a user-centred design approach and the targeted end-user is not required to learn how to use this system or remember or want to use it.

5.5   NON-CONTACT, VISION-BASED, CARDIOPULMONARY MONITORING (VCPM) SYSTEM

Contributors: Dr. Babak Taati and Michael Li, University Health Network—Toronto Rehabilitation Institute and Institute of Biomaterials and Biomedical Engineering, University of Toronto.

The non-contact, vision-based, cardiopulmonary monitoring system (VCPM) is a second example of a ZET in the application area of long-term prediction. The VCPM is used to monitor cardio-pulmonary activity of individuals while they sleep. This system does not require any device to be worn on the body and is being developed for the purpose of evaluating sleep apnea in individuals from neonates through to adults. The non-contact aspect of this system is important because current contact-based systems requiring monitoring devices to be worn on the body may not provide data that fully reflects habitual night time sleeping patterns of people in their own homes [110]. The current "gold standard" approach in sleep monitoring is the full-night supervised in-laboratory assessment with polysomnography (PSG) [111]. The PSG assessment involves recording various biological signals including electroencephalography (EEG), electromyography (EMG) of legs and chin, electrooculography (EOG), electrocardiography (ECG), respiratory effort, pulse oximetery and nasal pressure [111]. However, the high cost of conducting in-lab sleep assessments is a major problem that limits the application of PSG. Similarly, although portable sleep apnea monitoring devices would allow a person to sleep in his/her own bed, these devices still need to be worn on the body overnight. This is inconvenient for continuous monitoring over extended periods of time, and, therefore, ZETs capable of evaluating sleep apnea without contact would be ideal. Studies have shown respiratory failure often precedes sudden infant death syndrome and cardiac arrest, consequently, monitoring breathing disorders during sleep are especially important in neonates and older adults [112, 113]. In the younger population and teenagers, sleep apnea disorders are linked to higher neurobehavioural and cognitive disorders. Finally, in the adult population, sleep apnea is linked to a higher risk of developing hypertension [114]; cardiovascular disorders [115]; abnormalities in glucose metabolism [116]; and higher rates of car accidents [117]. On top of the social and emotional costs of sleep apnea, the economic costs of apnea in the United States alone are estimated to be $65–$165 billion annually [118]. Monitoring and the early diagnosis of sleep apnea could help in preventing its consequences and treating the patients before the disorder becomes more severe or results in irreversible health complications [114].

The VCPM uses computer vision techniques to capture the head and torso of the sleeping individual [110]. An IR light would be used and images would be sent to a remote computer for analysis and processing. See Figure 5.4 for an illustration of this setup.

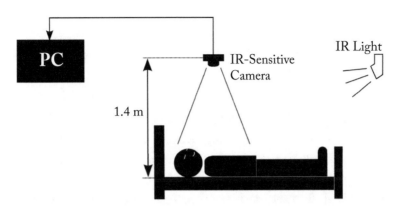

Figure 5.4: The VCPM system would consist of the IR light, an IR sensitive image sensor (camera), and a computer (not shown) to analyze and process the images (photo credit: University Health Network—Toronto Rehabilitation Institute).

Similar to the previous two examples, this device is also in an early stage of concept design and development. Previous research demonstrated that breathing rate could be accurately estimated (within 1 breath per minute, 97% of time) by way of processing near infrared (NIR) videos in various sleeping positions and even when a person is partially covered by a blanket sheet [110]. The VCPM was most recently trialed on seventeen adult participants (10 women and 7 men, average age 29.9 ± 11.7 years) in five simulated sleep positions. One position included an un-patterned white sheet for cover; the other four were without cover. Only one person was monitored at a time. In addition to respiration, heart rate was also monitored during sleep [111]. A mean percentage error of 3.4% and 5.0% was reported for respiratory rate and heart rate respectively [111]. For the data analysis, feature points were identified on captured IR video images and these points were tracked over time. The feature point trajectories were examined over time using optical flow to identify chest motion with respective to respiration and cardiac function. Blind source separation (i.e., principal or independent component analysis) was applied to trajectories to isolate possible signals of interest. Spectral analysis was used to characterize the periodicity and harmonics and this in turn would inform the optimal choice of signals for quantifying the heart rate and respiratory rate. In a study, for gold standard evaluation, heart rate was measured using a one lead ECG and respiration motion was measured using respiration belts placed around the chest and abdomen [111].

The next four ZET examples focus on applications in the area of Assistance with Daily Living Activities.

5.6   THE COACH

Contributor: Dr. Stephen Czarnuch, Department of Electrical and Computer Engineering, Memorial University of Newfoundland

The COACH (Cognitive Orthosis for Assisting with aCtivities in the Home) is an example of a ZET designed specifically to help older adults with cognitive impairments (e.g., Alzheimer's disease) through common activities of daily living (ADLs). COACH's developers chose to initially focus on one ADL, handwashing, to enable them to gain a good understanding of the context and to investigate and implement appropriate ways of modelling the problem and providing appropriate support to users [119]. As shown in Figure 5.5, the most current clinical version of COACH consists of a video camera mounted over the sink, a computer, speakers, and a flat-screen monitor [120]. The overall architecture of the system is illustrated in Figure 5.6. Using computer vision, COACH originally tracked the user's hands, the towel, and the soap as he or she interacted with the sink area. This was achieved by using image processing techniques that used different characteristics of the user's hand (e.g., skin colour), and the other relevant objects (e.g., location). More recently, a new hand tracker was developed to allow full 3D hand tracking independent of users' skin colour and texture, and also to reduce the system's sensitivity to lighting conditions [121]. This tracking information is passed to a planning module, which employs ML and inference techniques to determine autonomously where in the task the user is and to estimate parameters such as the user's overall level of dementia, current responsiveness, and preferred ordering of steps. The planning module then uses this information to decide the best course of action to take; namely, to continue to observe the user, give the user an audio or video prompt to guide him or her to the next step in the activity, or to summon the caregiver should the user require assistance.

COACH is a good example of a ZET as it does not require any explicit input from the user with dementia or the caregiver to make decisions and provide support autonomously. Importantly, the system follows the user through the task without the user having to wear any markers or other devices, avoiding possible non-compliance or annoyance from the person with dementia and also freeing the caregiver from needing to ensure the care recipient is wearing the tag or device, that it is charged, and that it is operational. COACH is able to learn about the user's preferences and is able to change its short- and long-term strategies to complement the dynamic nature of dementia. This includes the autonomous selection of appropriate prompts to match the user's abilities and current context, using prompts that range from simplistic (e.g., an audio prompt that says "Turn the water on") to specific (e.g., an audio prompt "John, push the silver handle to turn the water on" with an accompanying demonstrative video).

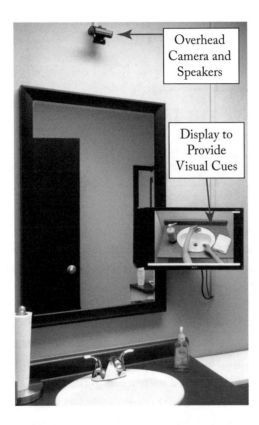

Figure 5.5: The components that make up an installed version of COACH—a webcam overhead to provide input to the system and a flat screen monitor and speakers to display the necessary prompts (photo credit: Intelligent Assistive Technology and Systems Lab, University of Toronto).

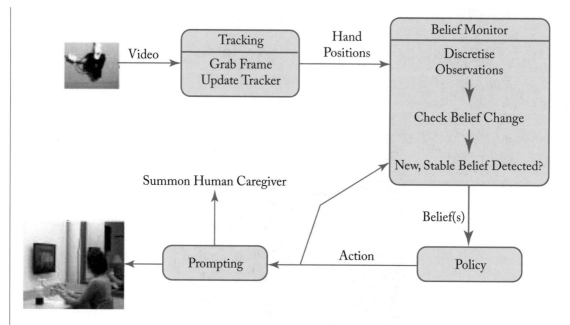

Figure 5.6: The overall architecture for the COACH system, which consists of the tracking, planning (belief monitor and policy), and prompting module (photo credit: Intelligent Assistive Technology and Systems Lab, University of Toronto).

In supervised pilot trials with COACH, participants with moderate to severe dementia were significantly more independent, requiring little or no human assistance. Participants with moderate-level dementia were able to complete an average of 11% more handwashing steps independently and required 60% fewer interactions with a human caregiver when COACH was in use. Four of the participants achieved complete or very close to complete independence [119]. In unsupervised trials with the same system, COACH struggled with hand tracking, identifying 46.6% of completed task steps, although the planning and prompting modules helped overcome many of the hand tracking limitations [120]. This unsupervised trial prompted the development of the new 3D hand tracker that was capable of correctly identifying over 99% of completed task steps [121]. Improvements are being made to fully integrate the new tracking module into the planning and prompting modules of COACH with the intention of deploying the device into people's homes for further real-world evaluation. In addition, the system is currently being developed for other ADLs, including tooth brushing, nutrition, and work-related tasks. Extension to additional ADLs typically involves a significant amount of time and expert knowledge. However, the use of a novel knowledge-driven method of rapidly specifying policies for use by the prompting module will allow COACH to be more easily configured to support these additional ADLs [122].

5.7   CULINARY ASSISTANT AND ENVIRONMENTAL SAFETY SYSTEM

Contributor: Dr. Hélène Pigot and Dr. Sylvain Giroux, DOMUS Laboratory, Faculty of Science, Université de Sherbrooke.

This next ZET example can be applied to three application areas: safety, assistance with daily activities of living, and communication. The culinary assistant and environmental safety system was developed to help support people living with cognitive impairment, especially from a traumatic brain injury, with autonomous meal preparation. This system was designed to be used in a supportive housing environment where the resident is not living alone. In Canada, each year, an estimated 100,000 individuals sustain a traumatic brain injury [123] and individuals with a severe traumatic brain injury often experience some degree of cognitive impairment [124]. This impairment may affect their individual judgement and awareness and consequently their ability to make decisions. As a result, preparing meals on their own can be difficult and dangerous [85, 125].

COOK, a culinary assistant and environmental safety system is composed of three integrated modules: activity recognition, cognitive assistance, and a distributed man-machine interface. It provides relevant cognitive assistance and safety information during meal preparation. Personal Assistant System (PAS), a preventative assistance system prompts the user, via audio, video, and visual methods, to take a specific corrective action in order to eliminate potentially hazardous situations. Security System (SS), the autonomous safety system will monitor the user's actions and identify potentially dangerous situations that have not been resolved. It can then "switch-off" power to a stove and inform care providers about a specific hazardous situation that has occurred [125]. To monitor the user's activity in the environment and to detect potentially dangerous situations (e.g., no one is monitoring the stove, while on), sensors are embedded throughout the environment, at the user's home. These sensors include, for example, motion, door, and window sensors; flow meters; temperature, humidity, and light level sensors; an intelligent stove; microphones; emergency monitoring sensors; touch screens, etc. [125]. In addition to sensors, outputs such as lights, speakers, computer screens/tablets, and mobile phones are used.

To use the culinary assistant COOK and its environmental safety system the user interacts with a touchscreen installed in close proximity to the stove. See the photo in Figure 5.7. The system is personalized for each user, according to his cognitive capacities and preferences. The user would select a pre-determined activity using the touchscreen interface, for example, the option to prepare a meal. The pre-configured system would then guide the user through the steps of preparing a meal using visual prompts on the touchscreen or other indicators mounted in the kitchen, such as LED lights that turn on to highlight items of interest. Figure 5.7 shows the user selecting to be monitored for recipes he already knows and needs only security follow up. Figure 5.8 shows how COOK guides the user to set the timer on the oven.

Figure 5.7: The main touchscreen user interface for COOK which allows a user to select the desired activity (photo credit: COOK video, Laboratoire DOMUS, Université de Sherbrooke).

Figure 5.8: Instructions to the user to set the oven timer (photo credit: COOK video, Laboratoire DOMUS, Université de Sherbrooke).

The COOK culinary assistant is the product of two previous cognitive assistant technologies: Archipel, developed to support people with mental disabilities [126], and SemAssist, developed to support people with semantic memory deficits [84, 127]. COOK models the cooking task according to the different steps that need to be completed and the associated tasks those steps correspond with. Any constraints that need to be adhered to, such as prerequisite steps that need to be completed, etc., are incorporated into this model [128, 129].

When the COOK is launched, the PAS and SS modules are also activated. User information is also used by the PAS and SS modules to set interaction preferences, specific risk management conditions, or specific trigger values (e.g., length of time user can leave stove unattended) [125]. During meal preparation, the embedded sensors monitor the user's action and the environment to identify potential risky situations (e.g., a user puts a pot of soup on a hotplate; then leaves the room to lie down leaving the soup on the hotplate unsupervised). When a potential risk situation or an error in execution is detected, the PAS initiates a personalized sequence of assistance interventions. It does this by following a set of pre-configured interventions itemized in a personalized and progressive preventative assistance tree (e.g., a series of specific, preferred actions, and communications to be followed for a user that may intensify as time increases). For each intervention, the user has three chances to take action before the next intervention on the tree is initiated (a Hierarchical Task Network preventative assistant tree structure) [125]. See Figure 5.9 for an example of a visual screen prompt.

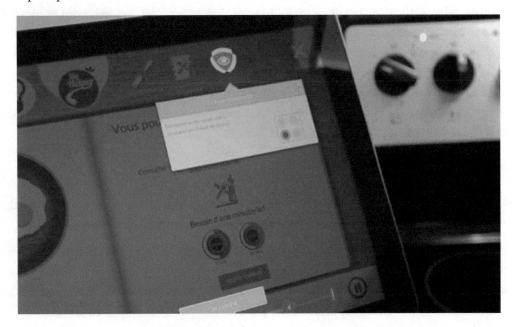

Figure 5.9: A warning prompt to the user informing him/her that the front left burner is on (photo credit: COOK video, Laboratoire DOMUS, Université de Sherbrooke).

If an intervention is successful, the PAS continues to monitor the user, otherwise the situation is unresolved and the SS locks COOK, turns off power to the stove, and notifies a care provider. The PAS module provides user prompts in five ways: jingles (short musical tunes/sounds), recorded voice messages, flashing lights, pictures, and/or text on the touchscreen. Jingles and voice messages can be varied based on three alert levels: remind, risk, and danger. Voice messages can be recorded with familiar voices. The PAS can also progressively increase the intensity of the communication prompts, for example, moving from implicit to explicit or indirect to direct prompts after a set time period and as risk level increases. An indirect explicit prompt example would be, "time to monitor your cooking?" and a direct explicit prompt example would be, "go monitor your cooking!"

Current research is examining COOK's ability to assist with meal preparation. The study is comparing four situations: (1) the user's ability to cook without COOK; (2) the user's ability to cook while learning the use of COOK; (3) the user's ability to cook while being supervised using COOK; and (4) the user using only the SS of COOK [84]. In addition, other methods will include interviews, observations, and questionnaires, only with the end-user and with care providers. The system in the study is customizable and is installed in a supportive housing complex. The system is being used daily by two residents with traumatic brain injury and the care providers supervising the residence (24 h/day and 7 d/week). A third participant will soon begin. This technology currently includes an interface that will also allow family caregivers to monitor the condition of the system user from a separate computer workstation [85]. Past studies have found that this system enabled two people living with traumatic brain injury to resume meal preparation safely with minimal assistance and this system may be expandable in future to other end-users, including the elderly [85].

Following user-design principles, the developers of the culinary assistant and environmental safety system have carefully considered the abilities and needs of the intended user group, using inputs, interaction modalities, and types of prompts that work best for this population. This system has been designed to prompt users only on an "as-needed" basis, and encourages the user to do as much as he or she can independently of the ZET.

5.8   PEAT

Contributor: Richard Levinson, CEO of BrainAid—Attention Control Systems Inc.; Acknowledgement: Dr. Henry Kautz, Goergen Institute for Data Science, University of Rochester.

Planning and Execution Assistant and Trainer (PEAT) is a ZET that runs on a cell phone and helps compensate for executive function impairment (i.e., the inability to remember a sequence of events or tasks), such as users with traumatic brain injury. Similar to the COOK Culinary Assistant and Environmental Safety System, PEAT provides assistance in the area of activities of daily living and incorporates prompting into the system. PEAT focuses on maintaining a schedule of a user's activities and automatically cueing the user when activities need to be started, resumed, or completed. A key aspect of PEAT is the use of reactive planning to adjust a user's schedule when an activity takes an unexpected amount of time to complete, or the user manually updates the calendar [130, 131]. PEAT represents each activity entered by the user as a task, each of which has temporal attributes such as a start time, end time, and expected duration. As the user progresses through their day and updates the status of tasks, PEAT reactively updates the day's schedule, and advises the user when tasks should start or stop, when conflicts arise, or when decisions must be made. Figure 5.10 shows an example of the PEAT interface and how it prompts a user for a current task.

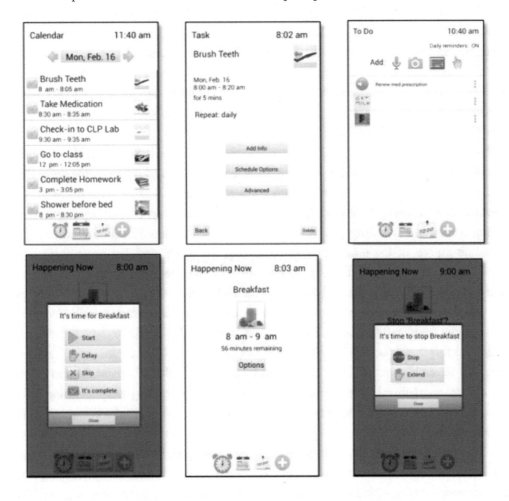

Figure 5.10: PEAT allows users to view tasks in calendar or individual task mode (see top row left and middle images). Users can add a task to their "To Do" list (see top right image). PEAT uses a unique display to present information about only the current activity happening now (see bottom middle image) and provides reminder cues to start, stop, and extend scheduled activities (see bottom left and right images) (photo credit: Brainaid.com).

PEAT uses both planning and activity recognition algorithms to schedule (or reschedule) activities as necessary. With respect to the planning algorithms, PEAT uses an AI planning system called PROgram Planning and Execution Language (PROPEL) [4, 130]. In PROPEL, users develop scripts for routine tasks such as activities required in the morning or going out shopping for groceries. The most important feature of PROPEL is that the same script can be used for both planning and execution. Planning involves simulating the script before it is executed. Scripts can contain choice points that identify where a choice must be selected from a set of alternative resources or subroutines. Examples of choice points in a "Dinner" script include choosing a restaurant for dinner and choosing between walking or driving. Without any planning a default script can be executed reactively by using heuristics to make default choice point selections. The planner first simulates the default program instance, and then simulates program variations. The planner evaluates each simulation with respect to the goals, and it searches for program variations that maximize goal achievement. Finally, the planner generates advice rules that are used during execution to make deliberate selections at choice points [130].

Research in this area has examined extending the AI of PEAT. Levinson, Halper, Harman, and Kautz (2009) [134] looked at the development of a Conversational Assistant for Rehabilitation (CARE) that included a CARE agent, therapy support, a conversational interface, context awareness, and machine learning. Chu et al. (2011) [132, 133] focused on developing a context-aware prompting system. Incorporating a context-aware system would allow PEAT to be knowledgeable about certain aspects of the user state and prompting would be modified accordingly to user's present situation. For example, if it is time for a user to take medication, he/she should not be prompted to do so if he/she is currently using the phone [133]. To develop a context-aware system, sensors would be needed to obtain further details about the user's current state. This sensor data may be obtained from motion sensors, contact sensors, or wearable RFID readers to detect a user's location, what they touch, or physiological state [133, 134]. In addition, explicit user queries could be used. This research examines an option-based hierarchical POMDP approach where different options can be programmed for selecting specific routines for prompting or questioning [133]. In addition, assistance is solicited from the user only when data from the sensors is ambiguous [133]. Simulation studies revealed that this approach could be used to deal with ambiguous situations while still successfully guiding the user through scheduled activities [132, 133]. Research studies are being conducted to evaluate this model, called "interactive activity recognition and prompting," in an android mobile device with patients with traumatic brain injury, post-traumatic stress disorder, pre-dementia or another cognitive impairment [133, 135].

The continued development of PEAT functionality, to improve machine intelligence and artificial intelligence, is an effort on the part of the developers to increase device usability.

5.9   BRAZE™ OBSTACLE DETECTION SYSTEMS

Contributor: Dr. Pooja Viswanathan, CEO of Braze Mobility and Intelligent Assistive Technology and Systems Laboratory, Department of Occupational Science and Occupational Therapy, University of Toronto.

Obstacle detection systems (ODS) mounted on wheelchairs are designed to improve the ability of wheelchair users to navigate within their environment and avoid collisions with obstacles in his/her pathway. Wheelchairs are often difficult to maneuver in tight and inaccessible spaces, leading to accidents, property damage, and loss of dignity. Visibility of obstacles in the rear, in particular, is limited by the backrests and/or headrests of the wheelchair, making backup in a wheelchair even more challenging for individuals with limited upper-body mobility and/or peripheral vision. While backup sensor technologies are now common in cars, these technologies have not existed for wheelchairs. AI use in these systems can range from ODS that: simply warn users of obstacles [136]; automatically halt wheelchairs to prevent an obstacle collision [136, 137]; alter the path of a wheelchair to avoid obstacles [136]; or are fully autonomous, taking a user from point A to B [136, 138, 139]. The wheelchairs requiring "zero effort" on the part of the wheelchair user tend to be those systems which either provide automatic feedback, stop the wheelchair to prevent collision, or divert a wheelchair from hitting an obstacle along its journey. In these situations, the ZETs would apply to the application area of activities of daily living, specifically mobility. Although there may be some minimal learning required when the systems are installed where the user learns what the wheelchair feedback means or why his/her wheelchair has stopped or has changed direction, the actual function of the wheelchair with respect to how it is used does not change. Fully autonomous wheelchairs, may require users to learn how to initially command the wheelchair, what commands are possible, and might also require the user to change how he/she operates the wheelchair (i.e., user just sits in the chair—no active control).

The ZET example presented here is the Braze™ ODS from Braze Mobility Inc., a new start-up company that recently launched its first ODS for wheelchairs. The Braze™ ODS is suitable for wheelchair users who are still cognitively capable of navigating their chairs independently and stopping their wheelchair before hitting an obstacle. This user feedback system was developed using a user-centred design approach and enhances the wheelchair user's awareness of his/her immediate surroundings, especially in places that are difficult to see such as behind the wheelchair or in his/her peripheral vision. In terms of functionality, when the ODS is setup and installed on a user's wheelchair (by user or therapist), the user uses his/her wheelchair the same as he/she typically does. When the wheelchair becomes too close to objects and the proximity of the wheelchair to the object falls within pre-defined "zones," the ODS automatically detects the obstacle and the user will receive feedback in the form of audible sounds, visual lights, and physical vibration.

Braze Mobility Inc. offers two types of ODS: the Sentina and Hydra. Both systems are low-cost, after-market products that can transform any wheelchair into a "smart" wheelchair. Ultrasonic sensors mounted on the wheelchair are able to detect obstacles. If the user's wheelchair begins to enter pre-programed "zones," the user is informed in several ways. An LED light display mounted on the front of the wheelchair will alight. The light display includes three lights, one corresponding to each direction: left, middle, and right. A yellow light warns the user that an object is in the "warning zone". A red light warns the user that he/she is in the "danger zone" (i.e., dangerously close to that object). For audio output, the ODS emits beeping sounds when it detects obstacles in the danger zone. Finally, the ODS will also create physical vibrations when an obstacle is in the pre-programed zones (i.e., weak vibration for warning zone, and strong vibration for danger zone). The systems can also be set to "short range" and "long range" modes, in order to detect obstacles both close by and further away. The ODS technology has been designed for easy installation without the need for tools and offers a high level of customizability. The Sentina product, shown in Figure 5.11, provides a fixed horizontal sensor coverage of 180° and vertical coverage of 50°, with an additional three sensors (Echo Heads™) with a 45° cone of coverage. In addition, there is the LED light display and control panel that indicates both obstacle location and proximity. The Hydra product, shown in Figure 5.12, does not include the fixed horizontal sensor but does include the three Echo Heads™, the LED display, and control panel [137]. With respect to power, both of these products can be powered via portable off-the-shelf power banks (USB connection) or via the wheelchair batteries.

This is a technology that has moved from research to the market place. Braze Mobility technologies were inspired by research conducted by Dr. Viswanathan and colleagues from the University of British Columbia; the Rehabilitation Sciences Institute, University of Toronto; and the University Health Network—Toronto Rehabilitation Institute. The Braze products were tested through a beta client program, in which over 30 users and caregivers in Canada and the U.S. provided feedback on the system. Current paying customers include U.S. veterans, and leading rehabilitation hospitals in Canada and the U.S. The products will soon be trialed in long-term care settings where elderly residents, especially those with cognitive impairment, are often excluded from the use of powered wheelchairs due to safety concerns. It is hoped that the Braze technology will widen access to safe and independent mobility. The founding team of Braze Mobility Inc. has conducted over a decade of research on smart wheelchair technologies and has published several user studies, see [136, 170–172] and www.iatsl.org/people/pviswanathan.html. This is a good example of a ZET that has matured over time with feedback from end-users via usability studies.

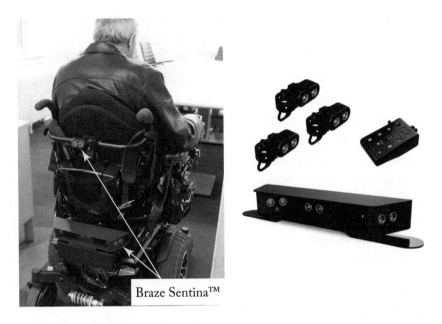

Figure 5.11: A photo of the Sentina ODS mounted on a wheelchair on left and the hardware components (photo credit: www.brazemobility.com).

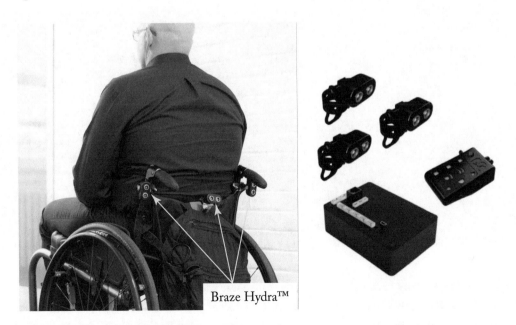

Figure 5.12: A photo of the Hydra ODS mounted on a wheelchair on left and the hardware components (photo credit: www.brazemobility.com).

5.10   THE HELPER

Contributor: Marge Coahran, University Health Network – Toronto Rehabilitation Institute.

The Health Evaluation and Logging Personal Emergency Response (HELPER) system is an example of a communication and monitoring ZET that is intended to support the safety and health of users in a home or institution. Following the ZET paradigm, the HELPER employs AI methods, such as computer vision, machine learning, and speech recognition, to automatically monitor a user in his or her home and detect adverse events, such as a fall, without the need for the user to initiate a call for help. Importantly, the device is very easy to use as it does not require occupants (users) to wear markers or modify their daily routines in any other way. A series of working prototypes that focus on fall detection have been designed, implemented, and tested in an ongoing iterative design improvement process [140–143]. Two versions of the system are shown in Figure 5.13, one intended for in-home use and the other designed for an institutional setting.

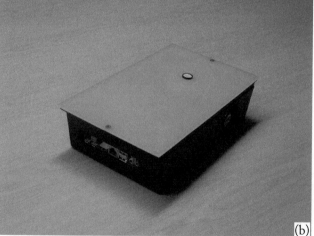

Figure 5.13: (a) An early HELPER prototype for in-home use that conducts a dialogue with the occupant to procure appropriate assistance. (b) A more recent HELPER prototype for institutional daytime or night-time use that sends alerts to smart phones carried by clinical staff (photo credit: Intelligent Assistive Technology and Systems, University of Toronto).

The system is a self-contained device that is installed on the ceiling in any room that requires monitoring. This approach maximizes coverage and minimises the possibility of falls being hidden from the device by furniture or other objects. The most recent version uses near-infrared illumination, invisible to the human eye, which allows it to operate equally well during the day or night without disturbing occupants in the room. The system generates video data and employs computer vision and video analysis techniques for background subtraction, to track the shapes of moving objects (humans, etc.) within the scene, and to extract a set of features that correlate with falls. ML techniques applied to these features allow the system to identify when a fall has likely occurred. An example of this process is illustrated in Figure 5.14.

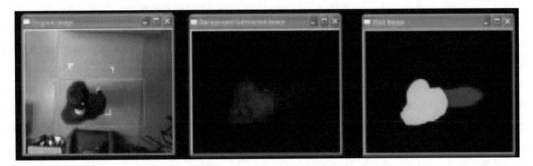

Figure 5.14: Example frames from the device, which illustrate the extraction of a person from the background and image analyses on the blob to separate the person from his or her shadow (photo credit: Intelligent Assistive Technology and Systems Laboratory, University of Toronto).

When a potential fall is detected, the system response differs based on the environment where the monitoring is taking place. The institutional version of the HELPER sends an alert message, including the room number and, optionally, an image of the room interior, via wifi to a set of associated smart phones carried by the appropriate clinical unit staff. This allows nurses to respond to falls immediately in situations where the fall is not heard, rather than discovering a fallen patient some time later during regular rounds. Note that the patients are not required to interact with the system in any way. In contrast, the home version of the system includes a speech-based interface (i.e., the user and HELPER device talk to each other) since the fallen person may be the only person present. In this version, when a potential fall is detected, the system initiates a conversation with the fallen person to determine if and what type of assistance he/she would like. A brief number of "yes/no" questions are asked, and automatic speech recognition is used to determine the user's responses. Typical questions include "Hello [person's name], do you need help?", "Do you want me to call an ambulance?", or "Do you want me to call your daughter?" If the person is unresponsive (e.g., he or she is unconscious or their speech unintelligible), the system defaults into connecting him/her to a live operator for further investigation. Moreover, as the system is speech-based, it is possible for the user to initialise a call for help even if the system cannot "see" him or her by calling out to the system. Further research is ongoing to improve the system's spoken-dialogue capabilities for handling emergency situations and the speech of older adult callers [144, 145].

It is important to note that the incoming video stream used by the system is not stored, but rather is analysed in real time and then discarded. Therefore, it is not possible for anyone to access past or live video streams and observe client activity. Moreover, the only data sent out by the system is the notification of an adverse event, either to nursing staff in a clinical setting, to a personal responder, or personal emergency response service, as determined through conversation between the user and the HELPER system. This approach not only saves on the transmission and storage of large amounts of data but also implicitly protects the person's privacy.

The HELPER has undergone testing in a variety of settings. The first prototype was tested extensively in the laboratory [141, 142] and later tested in two homes of healthy young adults who were instructed to simulate fall events several times daily for a week each [140, 143]. Although all simulated falls were detected by the second trial, many false alarms were also detected (e.g., from shadows, multiple occupants). The most recent prototype version includes the use of a wide-angle lens to increase the area of coverage, the use of near-infrared illumination to allow night-time operation and enhance robustness to abrupt lighting changes, and the ability to handle more than one subject by tracking multiple foreground regions. This system was tested in a 12-week study in 2 geriatric psychiatry units of a mental health hospital [146]. Study results demonstrated that the system could detect falls in a "real-world" situation; however, future work is needed to address the issue of false-alarms partially attributable to hospital infrastructure.

Note that the HELPER system is one example in a burgeoning field of automatic fall detection. Other technologies that are also being researched for this purpose include depth video from the Microsoft Kinect camera [147]; wearable gyroscopes and accelerometers, either as stand-alone sensors or internal to the user's smart phone [148]; and most recently, thermal cameras [149] and indoor radar systems [150]. To date, most evaluation studies still take place in laboratories, but as the technologies mature they are also tested in authentic environments like homes, long-term care facilities, and hospitals.

5.11   AMBIENT ACTIVITY TECHNOLOGIES

Contributors: Dr. Mark Chignell and Dr. Andrea Wilkinson, Interactive Media Lab, Mechanical and Industrial Engineering Department, University of Toronto.

Ambient Activity Technologies (AATs) are ZET devices developed to engage individuals with dementia in sensory play and exploration (cognitive stimulation) and to increase quality of life for these individuals, their families, and institutional staff. AATs are activity centres that may contain interactive visual displays, audio, and physical objects that can be touched and moved (e.g., switches, fabrics, pull handles, wheels, etc.). Most importantly, the technology does not require the user to learn how to use these devices in advance. AATs are being developed for many reasons: (1) to engage users and assist with responsive behaviour management (e.g., agitation, verbal abuse, hitting); (2) to provide users with an activity centre offering personalized content and experiences available at any time; (3) to provide users a way to keep active, exercise critical brain function, and maintain physical function; and (4) to offer an independent user activity that requires minimal caregiver and long-term care staff involvement or supervision [151]. The technology development follows a person-centered care approach that considers the multiple needs of the individual (e.g., biological, physical, social, and psychological). The AATs may be automatically personalized for a resident based on what the resident may choose to play with repeatedly or based on input from family members and staff (e.g., favorite music, interests, game difficulty level). The AAT will rely on body worn IDs (e.g., RFID), to recognize the system user and personalisation may be enhanced by keeping track of the user's play preferences. Two different AATs are presented here.

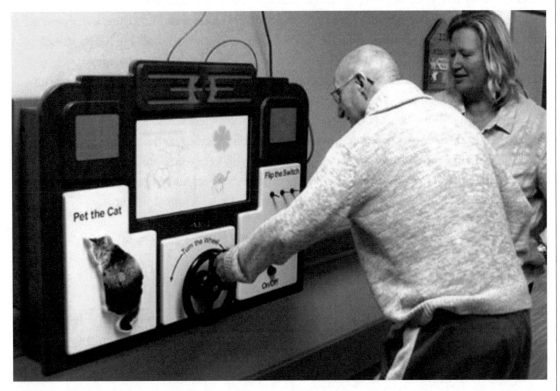

Figure 5.15: A patient at an Ontario long-term care facility interacting with Abby (credit: Ambient Activity Technologies).

Abby, is a wall-mounted activity centre that offers interactive activities such puzzles, games, and other challenges that will prompt users to touch, listen, and respond [152]. It is targeted towards individuals with later stages of dementia. The idea is to both accommodate and stimulate the person's remaining cognitive function and to engage the person in activity [152]. For example, Abby offers wheels that will activate a game on the video screen when they are turned, controls that can be moved (e.g., up and down switches, knobs that can be rotated clockwise or counterclockwise) to change video images, or to choose audio content (e.g., music for singing-along). There is also a cat with fluffy fur and if it is petted, videos of cats are shown. To use the system, the user simply walks up to the unit and begins to physically touch and move the objects. No setup is required on the part of the staff or the user and when the user walks away the unit remains as is until the next person comes to use it. See Figure 5.15 for a photo of an Abby unit in use in an Ontario long-term care home.

In 2017, Abby was evaluated (using a pre-test post-test design) in six Ontario long-term care homes to study how residents interact with the units and what outcomes were obtained after its use. Results reveal that responsive behaviours were significantly reduced and that some users were using the unit for extended periods of time [152]. The Abby units are now available for purchase, since December 2017, and have been installed in a number of Ontario long-term care homes.

The Centivizer system is also a wall-mounted activity centre but is focused more on reward-based games. It includes levers for pulling and buttons for pressing, similar to a slot machine at a casino with flashing lights and images. Similar to Abby, there is also a video screen for games, such as a Whack-A-Mole or Simon Says, and speakers for playing audio. This unit is targeted for individuals with early stage dementia, and may help with aging-in-place by supporting existing physical and cognitive function and focusing on muscles and cognitive functions required to perform activities of daily living [151, 152]. See Figure 5.16 for a picture of this unit.

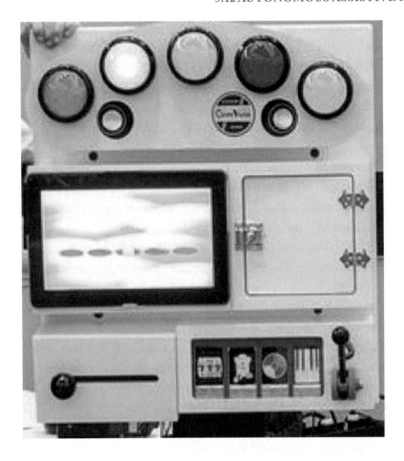

Figure 5.16: A prototype of the CentiVizer AAT (photo credit: Interactive Media Lab, University of Toronto).

These ZETs are currently undergoing usability studies to further improve their human-user interface design. They are using an interactive design process involving stakeholders such as designers, technologists, clinicians, caregivers, and end-users.

5.12   AUTONOMOUS ASSISTIVE ROBOTS

Contributors: Dr. Goldie Nejat, Autonomous Systems and Biomechatronics Laboratory (ASBLab), Mechanical and Industrial Engineering Department, University of Toronto.

Similar to the AAT and COOK, another ZET example in the application areas of leisure, cognitive stimulation, and activities of daily living are autonomous assistive robots. In this ZET example, two robots, named Tangy and Casper, developed in the ASBLab at the University of Toronto, will be described. Research studies have shown that engaging in group-based recreational activities can lead to benefits at the physical, social, emotional, and cognitive health levels [153]. Being engaged socially can help reduce the risk of dementia onset in older adults [154, 155]; improve the mental health of adolescents [156]; and the social networks of children [157]. With an increasing population of older adults, increasing burden on residential care staff in long-term care institutions, and a need for older adults to age-in-place (at home), robots that can engage older adults socially and/ or provide assistance would be beneficial for the older adult, residential care staff, and helping to support aging-in-place.

Tangy and Casper are intelligent, autonomous assistive robots, designed with the ability to reason about decisions, such as plan and schedule activities; determine their own appropriate assistive behaviors; detect users and objects in their environments; recognize naturally spoken language and non-verbal gestures and expressions; and respond using spoken-dialogue, facial expressions and gestures in order to effectively act in a given situation. The robots are designed with a number of onboard sensors that receive input from the user and environment, and communicate with the user through a range of output devices including speakers for speech, facial expressions, and head and body gestures using multi-degree-of-freedom joints, or visual display (e.g., videos or pictures on a tablet screen positioned on the robot). The robots were taught how to learn to recognize facial features (e.g., facial contours, eyebrows, eyes, nose, mouth) to identify different users [86]. The robots are also able to learn their assistive behaviors from non-experts using a 'teaching by demonstration' learning system architecture [158] and recognize a person's facial expressions [173] and body language [174] to estimate his/her emotional state. Users who interact with these robots do not need any training in advance and both robots have been developed following a user-centred design and empathy-based design (based on user's emotional state) approach.

Tangy is able to autonomously engage multiple users in group activities such as Bingo [160] and Trivia [161]. The tasks that Tangy needs to complete in order to facilitate a Bingo game include: (1) planning and scheduling of the games for multiple users throughout the day [162]; (2) locating individuals prior to the start of the group activity and reminding them to attend [163]; and (3) providing group-based and individualized assistance when needed during the group activity [160]). Tangy uses automated reasoning to plan and schedule group activities which has been tested using a multi-user system architecture [86]. Tangy was designed with a human-like upper torso that sits on top of a differential drive mobile platform [86]. An infrared sensor is used in the activity room where the Bingo game takes place. See Figure 5.17 for picture of a Tangy. As described in [160], Tangy responds in a synthesized voice; incorporates a 12" touchscreen tablet on its body to display Bingo numbers; can shake it's head (left-right) and nod (up-down); has two eyes that can move independently right-left, up-down, and together; can open and close its mouth; has two arms with movement (6° of freedom) in the shoulders, elbows, and wrists, and has grippers (can point). Tangy's sensors include two cameras in the eyes, one on top of the head, a laser range finder on a tilting platform for navigation, and an IR sensor that can track 3D joint locations to estimate a user's gestures. During user studies with older adults, Tangy's assistive behaviours follow finite state machines depending on the task (e.g., reminding someone or facilitating the bingo game). See [86] further information on the Bingo Game's finite state machine and an example of Tangy's behaviours during the game.

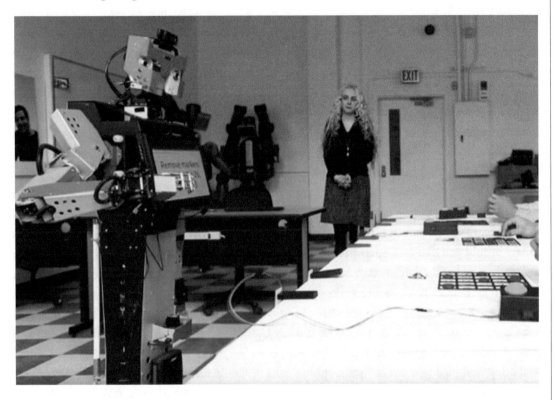

Figure 5.17: Picture of Tangy facilitating a Bingo game (photo credit: ASBLab, University of Toronto).

When Tangy is reminding people about an upcoming Bingo game, Tangy will navigate to the user's location based on his/her prior known schedule, identify the user based on facial recognition, and inform the user about the time and location of the Bingo game (e.g., verbal speech and using a visual on the tablet). When it is time to play the Bingo game, Tangy will go to the Bingo room and will help facilitate the game. Tangy greets the players, can call out Bingo numbers, and provides assistance to players. Assistance includes repeating the Bingo numbers already called, reviewing the players' cards, prompting a player if he/she has missed a called number or has incorrectly marked numbers, verifying winning Bingo cards, and celebrating Bingo wins. See Figure 5.17 to see Tangy in action. Tangy's performance in this task has been tested with 27 Bingo players with studies on-going [86, 160].

The robot Casper is designed to provide one-on-one cognitive stimulation and social engagement for people living with dementia [164]. Similar to Tangy, Casper accepts user input, provides assistive behaviors, and navigates the environment using sensor fusion, machine learning, and computer vision. Casper uses a laser scanner to help it navigate the environment, and uses 2D and 3D cameras to help it recognize users and objects (e.g., kitchen utensils). It has a microphone to detect user speech even in noisy environments and can communicate verbally. Casper is being designed to recognize certain emotions based on facial expressions, speech, and body language, and also learn based on the reactions and responses received by interacting with others. The robot has been designed to be sensitive enough to know when users may not want to engage socially and to adapt to the user's needs (e.g., cognitive abilities) [159]. Casper is a wheeled robot, approximately 4 feet tall, and unlike Tangy, has a face with LEDs for the eyebrows and the mouth to display different facial expressions. In its chest is a tablet for displaying visual information and accepting user input. See Figure 5.18 to see the robot's face and tablet screen.

Research studies with older adults are currently underway to "fine tune" Casper's abilities [164]. For example, during a meal planning task, Casper may look for a particular person, identify the person, make eye contact by waving, and say in a synthesized robotic voice, "Hi, how are you today? Would you like to prepare lunch together?" [164]. Casper would then escort the person to the kitchen and will ask if assistance is required in preparing the meal. If the answer is affirmative, Casper may demonstrate on the video screen various instructions, short videos, or guidance on how to prepare the meal. If the user has difficulty identifying or locating items, he/she can also ask Casper for assistance. When the meal is ready Casper may offer social comments such as, "that looks delicious," to suggest that it is time to eat [164]. Casper and other robots in the ASBLab are also being designed to assist in making phone calls to family members, engage the user in conversation, recommend clothing based on weather, and organize games and social activities [159, 165].

Casper was developed using off-the-shelf components which helps to keep costs for the robot to approximately $2,000. Casper is currently being targeted for use in private and long-term care homes, and retirement homes. Future work will examine how the robot will be able to also work with family members and caregivers to support aging-in-place. Other tasks being considered in the ASBLab for the assistive robots being developed may include "making beds, cleaning, cooking, and even supporting or lifting particularly frail individuals" [159]. See Figure 5.18 for a picture of Casper helping someone prepare a meal.

Figure 5.18: Casper interacts with a student by providing video instructions on how to make a tuna sandwich (photo credit: ASBLab, University of Toronto).

CHAPTER   6

Challenges and Future Directions

This book has provided an overview of the design, development, and testing of ZETs. In the previous section we discussed examples of different ZETs that could have a significant impact on the lives of people with disabilities. However, this field is still in its infancy. While considerable research and development efforts are underway, there are still limitations in this area of work that are preventing many ZETs from becoming widely available as commercial products.

6.1   CHALLENGES TO THE DEVELOPMENT OF ZETS

One limitation is the performance and robustness of the AI algorithms and sensing hardware. While these fields continue to make rapid progress, significant research and development needs to be conducted to ensure resulting systems are dependable and appropriate. For example, computer vision is becoming a more popular sensing modality in ZETs, however, many of the techniques being used are not robust enough to deal with real-world contexts and environments, such as changing lighting conditions within a room and multiple occupants in a house. Moreover, the guidelines and regulations regarding the capture and use of these types of data are in flux, which adds a layer of complexity to development and deployment.

Another limitation is the lack of real-world evidence that ZETs can have a significant positive effect on the lives of people with disabilities. While several ZETs have been developed, most of these have been evaluated in limited trials that often include only a small sample of subjects. In general, with early-stage technology development, it is extremely rare in the literature to find in-depth, long-term trials because of the associated costs and type of efficacy evaluated. As such, when compared with research studies in other healthcare fields, the strength of evidence for the use of ZETs is relatively low. However, much of this is a result of the newness of the field, namely the majority of technologies under development are just reaching the stages where they are ready to be trialed in real-world deployments for long periods of time. As the field matures and devices become more robust, larger, more robust trials will emerge.

This book also provided an overview of key design paradigms and approaches that can be used in the development of ZETs. However, there is no common framework, or guideline, for designers and researchers to follow. As a result, the ZET field is somewhat fragmented with different types of devices being developed that are not compatible, cannot be easily used together, and do not have efficacy outcomes that can be assessed and compared. While the principles of pervasive computing are often discussed in this field, they are not often adhered to. This is true with respect to ZETs under development as well as ZETs and assistive technologies that are commercially available. For example, as previously described, there is significant research being conducted on intelligent systems for powered wheelchairs, such as anti-collision and navigation systems. However, most of these devices are not compatible with the powered wheelchair that they were originally developed for and would require significant modifications to work with different chairs. This is why the Braze Mobility example is an example of a successful ZET as it has been designed to interface with any powered wheelchair. While work has been done to improve the plug-and-play capabilities of wheelchair accessories, accessing core functions, such as implementing a new drive-train controller, remains difficult. Most of these accessories are closed, proprietary systems. Agreed upon and adhered to frameworks that provide standardised practice for aspects such as communication protocols and other pervasive computing principles would help to alleviate many of these compatibility limitations.

Finally, there are practical limitations with current ZETs, namely issues around device installation and operation. As many of these devices are intended to be completely embedded into their environments, installation is a significant challenge especially when many different sensors and computing units need to be mounted on ceilings, walls, etc. For example, the fall detection system described in the case studies needs to be installed on the ceiling of each room that requires monitoring. These types of installations become even more difficult when retrofitting spaces that may not be conducive to supporting these technologies, such as an older home. Related to installation is the need to power devices and systems. As many ZETs require advanced processing and computing systems to operate, these devices cannot use batteries; they have power needs that require being hardwired or plugged into a power source. Using the fall detection system as an example once again, a limiting factor in the adoption of this type of technology is the fact that each ceiling mounted unit requires a power connection in the ceiling or power cables to be run to the nearest outlet. While potential solutions exist, such as retrofitting a ceiling light fixture with the ZET, many potential consumers of ZETs may not have this type of infrastructure available in their homes, nor would they want to run cables along ceilings and walls. A final functional limitation is that ZETs rely on infrastructure to operate, such as computer or wifi networks. This requires that the user already has or is willing to acquire the requisite infrastructure and that it can handle the needs of the ZET (e.g., adequate bandwidth and data streaming capabilities). The infrastructure must also be aligned with the purpose of the ZET. If, for example, a ZET monitored potentially life-threatening situations, such as fall detection, then the system should have more than one method of communicating with responders so that if one system fails, there is a back-up.

Finally, ZETs should be able to perform some degree of self-monitoring and failure detection; they need to have a way of letting people know when an error occurs or when they are offline. This is especially important with ZETs that are embedded into environments and interact with users if and when they are needed as the invisibility of these types of systems means the people using them may assume they are functional when they are not. Smart home systems are exploring ways of doing this, such as an email to designated users if a sensor goes offline or a central server loses connectivity with the system. However, new paradigms must be explored if these technologies are going to be used to support life-critical monitoring or be truly useful in supporting populations such as people living with early-stage dementia.

6.2   FUTURE CHALLENGES AND CONSIDERATIONS

Despite current limitations, the future of ZETs is promising and will likely continue to grow in use as the global population grows older and the number of people with one or more health conditions or disabilities increases. Regardless of the approach taken or components used, effective technologies must be adaptable to the needs, abilities, and preferences of the people that use them. Autonomous adaptability by the technology will become even more valuable as the trend towards providing care to people with disabilities in their own homes continues to grow. The ability to age in the place of one's choosing, at any age, is a win-win situation as this gives people more control over their own lives and enables care to be administered in a non-institutional setting, which can ease pressures on formal healthcare systems. Advanced sensing, ML, and other applications from the fields of pervasive computing and AI will continue to play a large role in the development of future ZETs, and indeed, technology in general. As people are not static nor homogeneous, it is apparent that ZETs must also take into account the whole context of the users it is intended to support, including abilities, environment, access to technology, and external support.

There are several new frontiers that are starting to be explored. One area is the use of advanced robotics to help care for people with a variety of disabilities. For example, personal robotics are being explored as a way to augment support for people who require care in their own homes. Robots can increasingly be able to assist with a variety of activities, such as meal preparation or house work, and in the future could also play a role in the monitoring and assessment of a person's overall health and wellbeing. Researchers are also investigating the placement of sensors into building materials. Dubbed *brick computing*, this approach creates pervasive computing systems that are made out of building materials themselves. Embedding a variety of sensors into floor, wall, and ceiling materials could enable continuous and autonomous ambient readings as the person moves about their home. For example, researchers at the University of Toronto have created a floor tile that is able to measure a person's physiological parameters (e.g., heart rate and blood pressure) by the person standing on the tile in bare feet. In this way, a person's physiological data can be captured several times a day in a home environment, allowing for a much richer, more accurate, long-term picture of the person's health. This significantly more holistic representation of health could be used to help inform new ways of detecting, monitoring, and treating potentially adverse health conditions as well as helping people understand and actively manage their own health.

The future of ZETs demands the creation of new design paradigms that will allow for data to be collected with respect to user needs. While traditional approaches, such as user-centred design, are useful, it is still difficult to truly involve representative users in the design process. This is especially true with vulnerable or frail users, such as older adults with dementia or other cognitive impairments. The dynamic nature of disabilities and their increasing prevalence is forcing this field to begin to incorporate design and evaluation strategies that are as equally dynamic. However, issues around how to incorporate these new approaches while still being able to construct prototypes in an efficient, timely, and cost-effective manner are significant barriers that need to be addressed.

As ZETs become more ubiquitous in healthcare institutions, communities, and homes, the social and ethical implications of these types of technologies must be addressed. The lack of explicit cognitive effort required to use a ZET demands that extra considerations need to be taken in order to ensure the users are consenting to using the technology, particularly when the targeted user may be unable to make an informed decision or consent about the ZET's use, such as people with cognitive disabilities and children. This issue is especially relevant for ZETs that support private and personal activities and are installed in sensitive locations (e.g., in the bathroom). Careful consideration needs to be taken in determining how consent will be obtained to use these new technologies, and how users will be educated about the potential benefits and limitations of these systems in a way that is consistent with their comprehension. As these issues have not been fully addressed within ZETs or pervasive technologies in general, it holds the potential for a new and fruitful area of research.

CHAPTER 7

Conclusions

While ZETs are a new concept, it embodies tremendous potential to provide customised and dynamic support with little or no effort on the part of the people who use them. Moreover, the rich and continuous data collected by ZETs coupled with the Internet of Things could enable new decision paradigms regarding interventions, which could, in turn, significantly improve people's health, quality of life, and overall wellbeing. Although they demonstrate exciting possibilities, the nature of ZETs also means that they must be carefully designed and tested, as a misuse of data or an error on the part of the ZET could have serious consequences. As such, ZETs must be developed in a way that holds the end users' wellbeing as the paramount goal and includes the stakeholders in the design process from start to finish ensuring the resulting technologies are appropriate, useful, and accepted.

Bibliography

[1]   J. Gubbi, R. Buyya, S. Marusic, and M. Palaniswami (2013). "Internet of Things (IoT): A vision, architectural elements, and future directions," *Future Generation Computer Systems*, 29(7) pp. 1645–1660, 2013/09/01. DOI: 10.1016/j.future.2013.01.010. 6, 11, 19

[2]   W. H. Organization (2002). "Towards a Common Language for Functioning, Disability and Health: ICF," World Health Organization, Geneva, Switzerland, Available: http://www.who.int/classifications/icf/training/icfbeginnersguide.pdf. 6, 19, 48

[3]   G. Abowd, M. Ebling, G. Hung, H. Lei, and H. W. Gellerson (2002). "Context-aware computing," *Pervasive Computing*, 1(3), pp. 22-23. DOI: 10.1109/MPRV.2002.1037718. 6, 34

[4]   J. E. Bardram (2004). "Applications of context-aware computng in hospital work: Examples and design principles," in *ACM Symposium on Applied Computing*, Nicosia, Cyprus, pp. 1574–1579: ACM. DOI: 10.1145/967900.968215. 6, 80

[5]   A. Dey (2001). "Understanding and using context," *Journal of Personal and Ubiquitous Computing*, 5(1), pp. 4–7. DOI: 10.1007/s007790170019. 6

[6]   C. Perera, A. Zaslavsky, P. Christen, and D. Georgakopoulos (2014). "Context aware computing for the Internet of Things: A survey," *IEEE Communications Surveys & Tutorials*, 16(1), pp. 414–454. DOI: 10.1109/SURV.2013.042313.00197. 6

[7]   J. E. Bardram, A. Mihailidis, and D. Wan (2007). *Pervasive Computing in Healthcare*. Boca Raton, FL: CRC Press. 7, 10

[8]   J. Paradiso (2011). "Guest editors' introcution: Smart energy systems," *IEEE Pervasive Computing*, 10(1), pp. 11–12. DOI: 10.1109/MPRV.2011.4. 7

[9]   Institute of Electrical and Electronic Engineers (IEEE) (2016). IEEE Smart Grid 2015 Annual Report. Available: https://smartgrid.ieee.org/images/files/pdf/2015_ieee_smart_grid_annual_report.pdf. 7, 8, 9

[10]   U. Hansmann, L. Merk, M. S. Nicklous, and T. Stober (2003). "Chapter 1: What pervasive computing is all about," in *Pervasive Computing*, 2nd Edition, Spring, Ed. New York: Spring-Verlag, pp. 11–22. 8, 9, 10, 14, 15

[11]   G. W. Arnold (2011). "Challenges and opportunities in smart grid: A position articile," *Proceedings of the IEEE*, 99(6), pp. 922–927. DOI: 10.1109/JPROC.2011.2125930. 9

[12]   V. Sundramoorthy (2011). "Domesticating energy-monitoring systems: Challenges and design concers," *IEEE Pervasive Computing*, 1(10), pp. 20–27. DOI: /10.1109/MPRV.2010.73. 9

[13]   D. Bergman, "Nonintrusive load-shed verification (2011), " *IEEE Pervasive Computing*, 1(10) pp. 49–57. DOI: 10.1109/MPRV.2010.71. 9, 10

[14]   U.S. Department of Commerce. (2014). NIST Framework and Roadmap for Smart Grid Interoperability Standards, Release 3.0, December Available: https://www.nist.gov/sites/default/files/documents/smartgrid/NIST-SP-1108r3.pdf. 10

[15]   M. Pollack and B. Peintner (2007). "Computer science tools and techniques," in *Pervasive Computing in Healthcare*, J. E. Bardram, A. Mihailidis, and D. Wan, Eds. Boca Raton, FL: CRC Press, pp. 21–40. 10, 11, 12, 13, 14, 15

[16]   F. A. Qayyum, M. Naeem, A. S. Khwaja, A. Anpalagan, L. Guan, and B. Venkatesh (2015). "Appliance scheduling optimization in smart home networks," *IEEE Access*, 3, pp. 2176–2190. DOI: 10.1109/ACCESS.2015.2496117. 14

[17]   A. Cavoukian (2011). "Privacy by design: The 7 foundational principle," Information and Privacy Commissioner of Ontario, Toronto. 15, 42

[18]   L. Chen, C. Nugent, J. Biswas, and J. Hoey (2011). *Activity Recognition in Pervasive Intelligent Environments*. London: Atlantis Press. DOI: 10.2991/978-94-91216-05-3. 16

[19]   L. Chen, J. Hoey, C. D. Nugent, D. J. Cook, and Z. Yu (2012). "Sensor-based activity recognition," *IEEE Transactions on Systems, Man, and Cybernetics, Part C (Applications and Reviews)*, 42(6) pp. 790–808. DOI: 10.1109/TSMCC.2012.2198883. 16

[20]   R. Szeliski (2010). *Computer Vision: Algorithms and Applications*. New York: Springer. 16, 19

[21]   G. Bradski and A. Kaehler (2008). *Learning OpenCV*. O'Reilly Media. 16, 19

[22]   R. Want (2006). *RFID Explained: A Primer on Radio Frequency Identification Technologies*. Morgan & Claypool Publishers. DOI: 10.2200/S00040ED1V01Y200602MPC001. 18

[23]   L. Shapiro and G. Stockman (2001). *Computer Vision*. Prentice Hall. 19

[24]   D. Forsyth and J. Ponce (2002). *Computer Vision: A Modern Approach*. Upper Saddle River, NJ: Prentice Hall. 19

[25]   R. O. Duda and P. E. Hart (2002). *Pattern Classification and Scene Analysis*. John Wiley and Sons. 19, 20

[26]   C. M. Bishop (2006). *Pattern Recognition and Machine Learning*. Springer. 19, 20, 22, 23, 27

[27]   M. Leo, G. Medioni, M. Trivedi, T. Kanade, and G. M. Farinella (2017). "Computer vision for assistive technologies," *Computer Vision and Image Understanding*, 154(Supplement C), pp. 1–15, 2017/01/01/. DOI: 10.1016/j.cviu.2016.09.001. 19

[28]   A. Krizhevsky, I. Sutskever, and G. E. Hinton (2012). "ImageNet classification with deep convolutional neural networks," presented at the *Advances in Neural Information Processing Systems 25 (NIPS 2012)*, Lake Tahoe, NV. 19

[29]   D. J. C. MacKay (2003). *Information Theory, Inference, and Learning Algorithms*. Cambridge University Press. 20, 26, 27, 32

[30]   C. Rasmussen (2001). "Joint likelihood methods for mitigating visual tracking disturbances," presented at the *Proceedings of the IEEE Workshop on Multi-Object Tracking*. DOI: 10.1109/MOT.2001.937983. 20

[31]   R. Sutton and A. G. Barto (1998). *Reinforcement Learning: An Introduction*. MIT Press. DOI: 10.1109/TNN.1998.712192. 20, 24

[32]   D. Poole and A. Mackworth (2010). *Artificial Intelligence: Foundations of Computational Agents*. Cambridge, UK: Cambridge University Press. DOI: 10.1017/CBO9780511794797. 20

[33]   S. Russell and P. Norvig (1995). *Artificial Intelligence: A Modern Approach*. Upper Saddle River, New Jersey: Prentice Hall. 20, 21

[34]   I. Goodfellow, Y. Bengio, and A. Courville (2016). *Deep Learning*. Cambridge, MA: MIT Press. 20

[35]   J. R. Quinlan (1993). *C4.5: Programs for Machine Learning*. San Mateo, CA: Morgan Kaufmann. 21, 22

[36]   Y. LeCun, Y. Bengio, and G. Hinton (2015). "Deep learning," *Nature*, 521, pp. 436–444, DOI: 10.1038/nature14539. 22

[37]   M. Markou and S. Singh (2003). "Novelty detection: A review - part 1: Statistical approaches," *Signal Processing*, 83, p. 2003. DOI: 10.1016/j.sigpro.2003.07.018. 22

[38]   C.-H. Teh and R. T. Chin (1988). "On image analysis by the methods of moments," *IEEE Transactions on Pattern Analysis and Machine Intelligence*, 10(4) pp. 496–513. DOI: 10.1109/34.3913. 23

[39]   T. Kohonen (1989). *Self-Organization and Associative Memory*. Berlin: Springer-Verlag. DOI: 10.1007/978-3-642-88163-3. 23

[40]   D. Koller and N. Friedman (2009). *Probabilistic Graphical Models: Principles and Techniques*. Cambridge, MA: MIT Press. 23, 26, 27

[41]   R. Bellman (1957). *Dynamic Programming*. Princeton, NJ: Princeton University Press. 24

[42]   M. L. Puterman (1994). *Markov Decision Processes: Discrete Stochastic Dynamic Programming*. New York: Wiley. DOI: 10.1002/9780470316887. 24

[43]   V. Mnih et al. (2015). "Human-level control through deep reinforcement learning," *Nature*, 518, p. 529–533. DOI: 10.1038/nature14236. 24

[44]   M. O. Duff (2002). *Optimal Learning: Computational procedures for Bayes-adaptive Markov Decision Processes*, University of Massachusetts, Amherst. 25

[45]   W. Pentney, M. Philipose, and J. Bilmes (2008). "Structure learning on large scale common sense statistical models of human state," presented at the *Proceedings of AAAI*, Chicago, July. 29

[46]   L. Chen, C. Nugent, M. Mulvenna, D. Finlay, X. Hong, and M. Poland (2008). "A logical framework for behaviour reasoning and assistance in a smart home," *International Journal of Assistve Robotics and Mechatronics*, 9(4), p. 2034. 29

[47]   F. Mastrogiovanni, A. Sgorbissa, and R. Zaccaria (2008). "An integrated approach to context specification and recognition in smart homes," in *Smart Homes and Health Telematics*, p. 2633. DOI: 10.1007/978-3-540-69916-3_4. 29

[48]   D. Salber, A. Dey, and G. Abowd (1999). "The context toolkit: Aiding the development of context-enabled applications," in *Conference on Human Factors in Computing Systems (CHI)*. DOI: 10.1145/302979.303126. 29

[49]   J. Hoey, T. Ploetz, D. Jackson, P. Olivier, A. F. Monk, and C. Pham (2011). "Rapid specification and automated generation of prompting systems to assist with people with dementia," in *Pervasive and Mobile Computing*. DOI: 10.1016/j.pmcj.2010.11.007. 29

[50]   H. Ryu and A. F. Monk (2009). "Interaction Unit Analysis: A new interaction design framework," *Human-Computer Interaction*, 24(4), pp. 367–407. DOI: 10.1080/07370020903038086?journalCode=hhci20#metrics-co. 29

[51]   L. Chen and P. Pu (2004). "A survey of preference elicitation methods," EPFL. 30

[52]   R. W. Picard (2000). *Affective Computing*. Cambridge, MA: MIT Press. 30

[53]   D. DeVault, R. Artstein, G. Benn, T. Dey, E. Fast, A. Gainer, K. Georgila, J. Gratch, A. Hartholt, M. Lhommet, G. Lucas, S. Marsella, F. Morbini, A. Nazarian, S. Scherer, G. Stratou, A. Suri, D. Traum, R. Wood, Y, Xu, A. Rizzo, and L.-P. Morency (2014). "SimSensei kiosk: a virtual human interviewer for healthcare decision support," presented at the *Proceedings of the 2014 International Conference on Autonomous Agents and Multi-agent Systems*, Paris, France. 31

[54] L. Lin, S. Czarnuch, A. Malhotra, L. Yu, T. Schröder, and J. Hoey (2014). "Affectively aligned cognitive assistance using Bayesian affect control theory," in *Ambient Assisted Living and Daily Activities: 6th International Work-Conference, IWAAL 2014*, Belfast, UK, December 2-5, 2014. Proceedings, L. Pecchia, L. L. Chen, C. Nugent, and J. Bravo, Eds. Cham: Springer International Publishing, pp. 279–287. DOI: 10.1007/978-3-319-13105-4_41. 31

[55] A. König, L. E. Francis, J. Joshi, J. M. Robillard, and J. Hoey (2017). "Qualitative study of affective identities in dementia patients for the design of cognitive assistive technologies," *Journal of Rehabilitation and Assistive Technologies Engineering*, 4, p. 2055668316685038. DOI: 10.1177/2055668316685038. 31

[56] M. Follette Story (1998). "Maximizing usabiity: The principle of universal design," *Assistive Technology*, 10, pp. 4–12. DOI: 10.1080/10400435.1998.10131955. 35

[57] D. A. Norman (2002). *The Design of Everyday Things*. New York: Basic Books. 37, 38

[58] J. D. Gould and C. Lewis (1985). "Designing for usability: Key principles and what designers think," *Communications of ACM*, 28(3), pp. 300–311. DOI: 10.1145/3166.3170. 37

[59] C. Wickens (1992). *Engineering Psychology and Human Performance*, 2nd ed. New York: Harper Collins Publishers, p. 560. 37

[60] J. Preece, Y. Rogers, and H. Sharp (2002). *Interaction Design*. Wiley & Sons. 38

[61] T. Adlam, R. Orpwood, and T. Dunn (2007). "User evaluation in pervasive healthcare," in *Pervasive Computing in Healthcare*, J. E. Bardram, A. Mihailidis, and D. Wan, Eds. Boca Raton, FL: CRC Press, pp. 243–274. 39

[62] K. Eason (1987). *Information Technology and Organizational Change*. London: Taylor & Francis. 39

[63] N. Dahlback, A. Jonsson, and L. Ahrenberg (1993). "Wizard of Oz studies: Why and how," in *IUI '93 Proceedings of the 1st International Conference on Intelligent User Interfaces*, New York: ACM. DOI: 10.1145/169891.169968. 40

[64] C. A. (2011). "Understanding empathy: Its features and effects," in *Empathy: Philosophical and Psychological Perspectives*, A. Coplan and P. Goldie, Eds.: Oxford University Press, pp. 3–18. 40

[65] A. Newell and P. Gregor (1997). "Human computer interaction for people with disabilities," in *Handbook of Human–Computer Interaction*, M. Helander, T. Landauer, and P. Prabhu, Eds. Amsterdam. 40

[66]   A. Newell, P. Gregor, and M. Morgan (2011). "User-sensitive inclusive design," *Universal Access in the Information Society*, 10,(3) pp. 235–243. DOI: 10.1007/s10209-010-0203-y. 40

[67]   S. Bødker (2006). "When second wave HCI meets third wave challenges," in *ACM 4th Nordic Conference on Human-Computer Interaction: Changing Roles*, pp. 1–8. DOI: 10.1145/1182475.1182476. 40

[68]   T. Jiancaro (in press). "Empathy-based design approaches," in *New Directions in 3rd Wave HCI: Volume 2 - Methodologies*, M. Filimowicz and V. Tzankova, Eds.: Springer. 40

[69]   A. Newell and P. Gregor (2000). "User sensitive inclusive design"—in search of a new paradigm," in *2000 Conf Univers Usability*, pp. 39–44. DOI: 10.1145/355460.355470. 40

[70]   A. Newell and A. Cairns (1993). "Designing for extraordinary users," *Ergon Des Q Hum Factors Appl*, 1(4) pp. 10–16. DOI: 10.1177/106480469300100405. 41

[71]   A. Newell and A. Carmichael (2006). "The use of theatre in requirements gathering and usability studies," *Interacting with Computers*, 18, pp. 996–1011. DOI: 10.1016/j.intcom.2006.05.003. 41

[72]   S. Lindsay, K. Brittain, D. Jackson, C. Ladha, K. Ladha, and P. Olivier (2012). "Empathy, participatory design and people with dementia," presented at the *SIGCHI Conference on Human Factors in Computing Systems*. DOI: 10.1145/2207676.2207749.

[73]   T. Kitwood (1997). "The experience of dementia," *Aging and Mental Health*, 1(1), pp. 13–22. DOI: 10.1080/13607869757344. 41

[74]   A. Mihailidis, J. Boger, and A. Cavoukian (2010). "Sensors and in-home collection of health data: A privacy by design approach," Information and Privacy Commissioner Ontario and Intelligent Assistive Technology and Systems Lab, Toronto, Canada. 41, 42, 44, 45, 46

[75]   J. F. Coughlin, L. A. D'Ambrosio, B. Reimer, and M. R. Pratt. (2007). "Older adult perceptions of smart home technologies: Implications for research, policy, and market innovations in healthcare," in *IEEE Proceedings of the Engineering in Medicine and Biology Confernece*, Lyon, France: IEEE. DOI: 10.1109/IEMBS.2007.4352665. 42

[76]   D. Kotz, S. Avancha, and A. Baxi (2009). "A privacy framework for mobile healt and homecare systems," in *SPIMACS'09*, Chicago, IL. DOI: 10.1145/1655084.1655086. 42

[77]   T. L. Beauchamp and J. F. Childress (2012). *Principles of Biomedical Ethics*, 7th ed. New York: Oxford University Press. 46

[78]   B. Friedman and J. Peter H. Kahn (2003). "Human values, ethics, and design," in *The Human-computer Interaction Handbook*, A. J. Julie and S. Andrew, Eds.: L. Erlbaum Associates Inc., pp. 1177–1201. 47

[79]   C. Shelley (2012). "Fairness in technological design," *Science and Engineering Ethics*, 18(4), pp. 663-680. DOI: 10.1007/s11948-011-9259-1. 47

[80]   M. Mulvenna, J. Boger, and R. Bond (2017). "Ethical by design - A manifesto," presented at the *35th European Conference on Cognitive Ergonomics (ECCE 2017)*, Umeå, Sweden, September 20–22, 2017. 47

[81]   J. C. Cornman, V. Freedman, and E. M. Agree (2005). "Measurement of assistive device use: Implications for estimates of device use and disability in later life," *Gerontologist*, 45(3), p. 347. DOI: 10.1093/geront/45.3.347. 48

[82]   M. J. Scherer, T. Hart, N. Kirsch, and M. Schulthesis (2005). "Assistive technologies for cognitive disabilities.," *Critical Reviews in Physical and Rehabilitation Medicine*, 17(3), p. 195. DOI: 10.1615/CritRevPhysRehabilMed.v17.i3.30. 48

[83]   R. S. Wilson, C. F. Mendes De Leon, L. L. Barnes, J. A. Schneider, J. L. Bienias, D. A. Evans, and D. A. Bennett. (2002). "Participation in cognitively stimulating activities and risk of incident Alzheimer Disease," *Journal of the American Medical Association*, 287(6) pp. 742–748, February. 49

[84]   S. Giroux, N. Bier, H. Pigot, B. Bouchard, A. Bouzouane, M. Levasseur, M. Couture, C. Bottari, B. Swaine, P.-Y. Therriault, K. Bouchard, F. Le Morellec, S. Pinard, S. Azzi, M. Olivares, T. Zayani, G. LeDorze, P. De Loor, A. Thepaut, and B. Le Pevedic (2015). "Cognitive assistance to meal preparation: design, implementation, and assessment in a living lab," in *2015 AAAI Spring Symposium Series*, pp. 01–25. 50, 77, 78

[85]   L. DOMUS. (2017). COOK culinary assistant. Available: https://www.usherbrooke.ca/domus/en/research/research-projects/cook-culinary-assistant/. 50, 75, 78

[86]   W.-Y. G. Louie, T. Vaquero, G. Nejat, and J. C. Beck (2014). "An autonomous assistive robot for planning, scheduling and facilitating multi-user activities," in *IEEE International Conference on Robotics and Automation (ICRA)*, pp. 5292–5298: IEEE. DOI: 10.1109/ICRA.2014.6907637. 50, 90, 91

[87]   L. C. Watson, C. L. Lewis, C. G. Moore, and D. V. Jeste (2010). "Perceptions of depression among dementia caregivers: findings from the CATIE-AD trial," *International Journal of Geriatric Psychiatry*, 26(4), pp. 397–402. DOI: 10.1002/gps.2539. 51

[88]   R. Schulz and S. Beach (1999). "Caregiving as a risk factor for mortality: The caregiver health effects study," *Journal of the American Medical Association*, 282(23), pp. 2215–2219. DOI: 10.1001/jama.282.23.2215. 51

[89]   S. Bouakaz, M. Vacher, M.-E. Bobillier Chaumon, F. Aman, S. Bekkadja, F. Portet, E. Guillou, S. Rossato, E. Desseree, P. Traineau, J.-P. Vimont, and T. Chevalier (2014). "CIRDO: Smart companion for helping elderly to live at home for longer," *Innovation and Research in Biomedical Engineering*, 35(2), pp. 100–108, 2014/04/01. DOI: 10.1016/j. irbm.2014.02.011. 55

[90]   M. Chan, D. Esteve, C. Escriba, and E. Campo (2008). "A review of smart homes- present state and future challenges," *Computer Methods and Programs in Biomedicine*, 91(1), pp. 55–81. DOI: 10.1016/j.cmpb.2008.02.001. 55

[91]   B. Gillett, B. Peckler, R. Sinert, C. Onkst, S. Nabors, S. Issley, C. Maguire, S. Galwankarm, and B. Arquilla (2008). "Simulation in a disaster drill: Comparison of high-fidelity simulators versus trained actors," *Academic Emergency Medicine*, 15(11), pp. 1144–1151. DOI: 10.1111/j.1553-2712.2008.00198.x. 55

[92]   L. Ten Bosch (2003). "Emotions, speech and the ASR framework," *Speech Communication*, 40(1), pp. 213–225. DOI: 10.1016/S0167-6393(02)00083-3. 55

[93]   N. Campbell (2000). "Databases of emotional speech," in *ISCA Tutorial and Research Workshop (ITRW) on Speech and Emotion: Developing a Conceptual Framework*, Newark, Northern Ireland, pp. 34-39: International Speech Communication Association. 55

[94]   V. Lockton, A. Mihailidis, J. Boger, and M. Chibba (2010). *Sensors and In-Home Collection of Health Data: A Privacy by Design Approach*. Information and Privacy Commissioner, Ontario, Canada. 56

[95]   A. J. Bharucha, V. Anand, J. Forlizzi, M. A. Dew, C. F. Reynolds, S. Stevens, and H. Wactlar (2009). "Intelligent assistive technology applications to dementia care: Current capabilities, limitations, and future challenges," *American Journal Geriatric Psychiatry*, 17(2), pp. 88–104. DOI: 10.1097/JGP.0b013e318187dde5. 59

[96]   A. M. Cook and J. M. Polgar (2015). *Assistive Technologies: Principles and Practice*, 4th ed. St. Louis, MO: Elsevier. 59

[97]   B. O'Neill and A. Gillespie (2014). *Assistive Technology for Cognition: A Handbook for Clinicians and Developers*. Psychology Press. 59

[98]   M. Ienca, J. Fabrice, B. Elger, M. Caon, A. S. Pappagallo, R. W. Kressig, and T. Wangmo (2017). "Intelligent assistive technology for Alzheimer's Disease and other dementias: A

systematic review," *Journal of Alzheimer's Disease*, 56(4), pp. 1301–1340. DOI: 10.3233/ JAD-161037. 59

[99] WHO (2017). "Dementia Fact Sheet," World Health Organization, December 2017, Available: http://www.who.int/mediacentre/factsheets/fs362/en/, Accessed: December 10, 2017. 65

[100] CIHI (2017). "Caring for seniors with Alzheimer's Disease and other forms of dementia," 2010. Available: https://secure.cihi.ca/free_products/Dementia_AIB_2010_EN.pdf, Accessed: December 7, 2017. 65

[101] A. Bankole et al. (2011). "Continuous, non-invasive assessment of agitation in dementia using inertial body sensors," in *Proceedings of the 2nd Conference on Wireless Health*, San Diego, CA, pp. 1–9: ACM. DOI: 10.1145/2077546.2077548. 65

[102] J. Cerejeira, L. Lagarto, and E. B. Mukaetova-Ladinska (2012). "Behavioral and psychological symptoms of dementia," *Frontiers in Neurology*, 3, p. 73. DOI: 10.3389/ fneur.2012.00073. 65

[103] ASC. (December 7, 2017,). "Dementia numbers in Canada." Available: http://www.alzheimer.ca/en/about-dementia/what-is-dementia/dementia-numbers. 65

[104] K. Pemberton (2016). "Danger in the dementia care home," in *The Vancouver Sun*, http:// www.vancouversun.com/health/danger+dementia+care+home/11655805/story.html. 65

[105] TLFP (2016). "Wave of dementia behind 12 homicides has Ontario nursing homes pleading for help from province," in *National Post*, ed: The London Free Press. 65

[106] S. S. Khan, T. Zhu, B. Ye, A. Mihailidis, A. Iaboni, K. Newman, A. H. Wang, and L. Schindel Martin (2017). "DAAD: A framework for detecting agitation and aggression in people living with dementia using a novel multi-modal sensor network," presented at the *International Conference on Data Mining*, New Orleans. DOI: 10.1109/ICDMW.2017.98. 65, 67

[107] S. I. Chaudhry, J. A. Mattera, J. P. Curtis, J. A. Spertus, J. Herrin, Z. Lin, C. Phillips, B. V. Hodshon, L. S. Cooper, and H. M. Krumholz (2010). "Telemonitoring in patients with heart failure," *New England Journal of Medicine*, 363(24), pp. 2301–2309. DOI: 10.1056/ NEJMoa1010029. 67

[108] M. Alwan (2009). "Passive in-home health and wellness monitoring: Overview, value and examples," in *Engineering in Medicine and Biology Society, 2009. EMBC 2009. Annual International Conference of the IEEE*, pp. 4307–4310: IEEE. DOI: 10.1109/ IEMBS.2009.5333799. 67

[109]   S. L. Grace, G. Taherzadeh, J. Chang, J. Boger, A. Arcelus, S. Mak, C. Chessex, and A. Mihailidis (2017). "Perceptions of seniors with heart failure regarding autonomous zero-effort monitoring of physiological parameters in the smart-home environment," *Heart and Lung: The Journal of Acute and Critical Care*, 46(4), pp. 313–319. DOI: 10.1016/j.hrtlng.2017.04.007. 68

[110]   M. H. Li, A. Yadollahi, and B. Taati (2014). "A non-contact vision-based system for respiratory rate estimation," in *Engineering in Medicine and Biology Society (EMBC), 2014 36th Annual International Conference of the IEEE*, pp. 2119–2122: IEEE. DOI: 10.1109/EMBC.2014.6944035. 70, 71

[111]   M. H. Li, A. Yadollahi, and B. Taati (2017). "Non-contact vision-based cardiopulmonary monitoring in different sleeping positions," *IEEE Journal of Biomedical and Health Informatics*, 21(5), pp. 1367–1375. DOI: 10.1109/JBHI.2016.2567298. 70, 71

[112]   F. Al-Khalidi, R. Saatchi, H. Elphick, and D. Burke (2011). "An evaluation of thermal imaging based respiration rate monitoring in children," *American Journal of Engineering and Applied Sciences*, 4(4), pp. 586–597. DOI: 10.3844/ajeassp.2011.586.597. 70

[113]   J. Rosenberg, M. Pedersen, T. Ramsing, and H. Kehlet (1992). "Circadian variation in unexpected postoperative death," *British Journal of Surgery*, 79(12), pp. 1300–1302. DOI: 10.1002/bjs.1800791219. 70

[114]   P. E. Peppard, T. Young, M. Palta, and J. Skatrud (2000). "Prospective study of the association between sleep-disordered breathing and hypertension," *New England Journal of Medicine*, 342(19), pp. 1378–1384. DOI: 10.1056/NEJM200005113421901. 70

[115]   Y. Peker, J. Carlson, and J. Hedner (2006). "Increased incidence of coronary artery disease in sleep apnoea: a long-term follow-up," *European Respiratory Journal*, 28(3), pp. 596–602. DOI: 10.1183/09031936.06.00107805. 70

[116]   N. M. Punjabi and V. Y. Polotsky (2005). "Disorders of glucose metabolism in sleep apnea," *Journal of Applied Physiology*, 99(5), pp. 1998–2007. DOI: 10.1152/japplphysiol.00695.2005. 70

[117]   J. Teran-Santos, A. Jimenez-Gomez, J. Cordero-Guevara, and C. G. Burgos–Santander (1999). "The association between sleep apnea and the risk of traffic accidents," *New England Journal of Medicine*, 340(11), pp. 847–851. DOI: 10.1056/NEJM199903183401104. 70

[118]   McKinsey and Company - Harvard Medical School (2010). "The price of fatigue: The surprising economic costs of unmanaged sleep apnea," Harvard Medical School, Available: https://sleep.med.harvard.edu/file_download/100. 70

[119]   A. Mihailidis, J. Boger, M. Candido, and J. Hoey (2008). "The COACH prompting system to assist older adults with dementia through handwashing: An efficacy study," *BMC Geriatrics*, 8(28). DOI: 10.1186/1471-2318-8-28. 72, 74

[120]   S. Czarnuch, S. Cohen, V. Parameswaran, and A. Mihailidis (2013). "A real-world deployment of the COACH prompting system," *Journal of Ambient Intelligence and Smart Environments*, 5(5), pp. 463–478. 72, 74

[121]   S. Czarnuch and A. Mihailidis (2016). "Development and evaluation of a hand tracker using depth images captured from an overhead perspective," *Disability and Rehabilitation: Assistive Technology*, 11(2), pp. 150–157. DOI: 10.3109/17483107.2015.1027304. 72, 74

[122]   M. Grześ, J. Hoey, S. Khan, A. Mihailidis, S. Czarnuch, D. Jackson, and A. Monk (2014). "Relational approach to knowledge engineering for POMDP-based assistance systems as a translation of a psychological model," *International Journal of Approximate Reasoning*, 55(1), pp. 36–58. DOI: 10.1016/j.ijar.2013.03.006. 74

[123]   J.-M. Billette and T. Janz (2011). "Les blessures au Canada: Un aperçu des résultats de l'Enquête sur la santé dans les collectivités canadiennes," *Statistique Canada*, no. 82–624. 75

[124]   C. Bottari, C. Dassa, C. Rainville, and É. Dutil (2010). "A generalizability study of the instrumental activities of daily living profile," Archives of *Physical Medicine and Rehabilitation*, 91(5), pp. 734–742. DOI: 10.1016/j.apmr.2009.12.023. 75

[125]   M. Olivares, S. Giroux, P. De Loor, A. Thepaut, H. Pigot, S. Pinard, C. Bottar, G. Le Dorze, and N. Bier (2016). "An ontology model for a context-aware preventive assistance system: reducing exposition of individuals with Traumatic Brain Injury to dangerous situations during meal preparation." *2nd IET International Conference on Technologies for Active and Assisted Living (TechAAL 2016)*. DOI: 10.1049/ic.2016.0052. 75, 77

[126]   J. Bauchet, H. Pigot, S. Giroux, D. Lussier-Desrochers, Y. Lachapelle, and M. Mokhtari (2009). "Designing judicious interactions for cognitive assistance: the acts of assistance approach," in *Proceedings of the 11th International ACM SIGACCESS Conference on Computers and Accessibility*, pp. 11–18: ACM. DOI: 10.1145/1639642.1639647. 77

[127]   N. Bier, J. Macoir, S. Joubert, C. Bottari, C. Chayer, H. Pigot, S. Giroux, and SemAssist Team (2011). "Cooking "Shrimp à la Créole": A pilot study of an ecological rehabilitation in semantic dementia," *Neuropsychological Rehabilitation*, 21(4), pp. 455–483. DOI: 10.1080/09602011.2011.580614. 77

[128]   J. Bauchet, S. Giroux, H. Pigot, D. Lussier-Desrochers, and Y. Lachapelle (2008). "Pervasive assistance in smart homes for people with intellectual disabilities: A case study on

meal preparation.," *International Journal of Assistive Robotics and Mechatronics (IJARM)*, 9(4), pp. 42–54. 77

[129] M. Najjar, F. Courtemanche, H. Hamam, A. Dion, and J. Bauchet (2009). "Intelligent recognition of activity of daily living for assisting memory and/or cognitively impaired elders in smart homes," *International Journal of Ambient Computing and Intelligence (IJACI)*, 1(4), pp. 46–62. DOI: 10.4018/jaci.2009062204. 77

[130] R. Levinson (1995). "A general programming language for unified planning and control," *Artificial Intelligence: Special Issue on Planning and Scheduling*, vol. 76. 79, 80

[131] J. Modayil, R. Levinson, C. Harman, D. Halper, and H. Kautz (2008). "Integrating sensing and cueing for more effective activity reminders," presented at the *Proceedings AAAI Fall 2008 Symposium on AI in Eldercare: New Solutions to Old Problems*, Washington, DC, November, 2008. 79

[132] Y. Chu, Y. Chol Song, H. Kautz, and R. Levinson (2011). "When did you start doing that thing that you do," in *Interactive Activity Recognition and Prompting. AAAI 2011 Workshop on Artificial Intelligence and Smarter Living*. 80

[133] Y. Chu, Y. Chol Song, R. Levinson, and H. Kautz (2012). "Interactive activity recognition and prompting to assist people with cognitive disabilities," *Journal of Ambient Intelligence and Smart Environments*, 4(5), pp. 443–459. 80

[134] R. Levinson, D. Halper, H. Kautz, and C. Harman (2009). "A conversational cognitive aid with activity monitoring, planning and execution," in *International Joint Conference on Artificial Intelligence (IJCAI), Workshop on Intelligent Systems for Assisted Cognition*, Pasadena, CA. 80

[135] BrainAid. (December 7, 2017). "BrainAid." Available: http://brainaid.com/. 80

[136] P. Viswanathan, E. P. Zambalde, G. Foley, J. L. Graham, R. H. Wang, B. Adhikari, A. K. Mackworth, A Mihailidis, W. C. Miller, and I. M. Mitchell (2017). "Intelligent wheelchair control strategies for older adults with cognitive impairment: user attitudes, needs, and preferences," *Autonomous Robots*, 41(3), pp. 539–554. DOI: 10.1007/s10514-016-9568-y. 81, 82

[137] R. Wang, A. Mihailidis, T. Dutta, and G. Fernie (2011). "Usability testing of multimodal feedback interface and simulated collision-avoidance power wheelchair for long-term-care home residents with cognitive impairments," *Journal of Rehabilitation Research and Development*, 48(7), pp. 801–822. DOI: 10.1682/JRRD.2010.08.0147. 81, 82

[138]   M. Scudellari (2017). "Self-driving wheelchairs debut in hospitals and airports," vol. 2017, ed: *IEEE Spectrum, Biomedical Engineering Blog.* DOI: 10.1109/MSPEC.2017.8048827. 81

[139]   J. Pineau, A. K. Moghaddam, H. K. Yuen, P. S. Archambault, F. Routhier, F. Michaud, and P. Boissy (2014). "Automatic detection and classification of unsafe events during power wheelchair use," *IEEE Journal of Translational Engineering in Health and Medicine,* 2, pp. 1–9. DOI: 10.1109/JTEHM.2014.2365773. 81

[140]   M. Belshaw, B. Taati, J. Snoek, and A. Mihailidis (2011). "Towards a single sensor passive solution for automated fall detection," in *Engineering in Medicine and Biology Society, EMBC, 2011 Annual International Conference of the IEEE,* pp. 1773–1776: IEEE. DOI: 10.1109/IEMBS.2011.6090506. 84

[141]   T. Lee and A. Mihailidis (2005). "An intelligent emergency response system: preliminary development and testing of automated fall detection," *Journal of Telemedicine and Telecare,* 11(4) pp. 194–198. DOI: 10.1258/1357633054068946. 84, 86

[142]   M. Hamill, V. Young, J. Boger, and A. Mihailidis (2009). "Development of an automated speech recognition interface for personal emergency response systems," *Journal of Neuro-Engineering and Rehabilitation,* 6(1), p. 26. DOI: 10.1186/1743-0003-6-26. 84, 86

[143]   M. Belshaw, B. Taati, D. Giesbrecht, and A. Mihailidis (2011). "Intelligent vision-based fall detection system: preliminary results from a real-world deployment," *RESNA/ICTA,* pp. 5–8. 84, 86

[144]   V. Young and A. Mihailidis (2013). "The CARES corpus: a database of older adult actor simulated emergency dialogue for developing a personal emergency response system," *International Journal of Speech Technology,* 16(1), pp. 55–73. DOI: 10.1007/s10772-012-9157-1. 86

[145]   V. Young, E. Rochon, and A. Mihailidis (2016). "Exploratory analysis of real personal emergency response call conversations: considerations for personal emergency response spoken dialogue systems," *Journal of Neuroengineering and Rehabilitation,* 13(1), p. 97. DOI: 10.1186/s12984-016-0207-9. 86

[146]   M. Coahran et al. (2018). "Evaluation of an automated fall detection technology in inpatient geriatric psychiatry: Nurses perceptions and key lessons learned," *Canadian Journal on Aging,* 37(3). 86

[147]   E. E. Stone and M. Skubic (2015). "Fall detection in homes of older adults using the Microsoft Kinect," *IEEE Journal of Biomedical and Health Informatics,* 19(1), pp. 290–301. DOI: 10.1109/JBHI.2014.2312180. 86

[148]   M. Kangas, R. Korpelainen, I. Vikman, L. Nyberg, and T. Jämsä (2015). "Sensitivity and false alarm rate of a fall sensor in long-term fall detection in the elderly," *Gerontology*, 61(1), pp. 61–68. DOI: 10.1159/000362720. 86

[149]   J. Rafferty, J. Synnott, C. Nugent, G. Morrison, and E. Tamburini (2016). "Fall detection through thermal vision sensing," in *Ubiquitous Computing and Ambient Intelligence: 10th International Conference, UCAmI 2016*, San Bartolomé de Tirajana, Gran Canaria, Spain, November 29–December 2, 2016, Part II 10, pp. 84–90: Springer. DOI: 10.1007/978-3-319-48799-1_10. 86

[150]   L. Liu, M. Popescu, M. Skubic, M. Rantz, and P. Cuddihy (2016). "An automatic in-home fall detection system using Doppler radar signatures," *Journal of Ambient Intelligence and Smart Environments*, 8(4), pp. 453–466. DOI: 10.3233/AIS-160388. 86

[151]   A. Wilkinson, V. Charoenkitkarn, J. O'Neill, M. Kanik, and M. Chignell (2017). "Journeys to engagement: Ambient activity technologies for people living with dementia," in *Proceedings of the 26th International Conference on World Wide Web Companion*, pp. 1103–1110: International World Wide Web Conferences Steering Committee. DOI: 10.1145/3041021.3054933. 87, 88

[152]   University of Toronto. (November 29, 2017). "Engineering researcher develops technologies to reduce problem behaviours in people with dementia|Department of Mechanical and Industrial Engineering." Available: http://www.mie.utoronto.ca/u-of-t-engineer-develops-technologies-to-reduce-problem-behaviours-in-people-with-dementia/. 88

[153]   L. L. Caldwell (2005). "Leisure and health: why is leisure therapeutic?," *British Journal of Guidance & Counselling*, 33(1), pp. 7–26. DOI: 10.1080/03069880412331335939. 90

[154]   J. Verghese et al. (2003). "Leisure activities and the risk of dementia in the elderly," *New England Journal of Medicine*, 348(25), pp. 2508–2516. DOI: 10.1056/NEJMoa022252. 90

[155]   M. Crowe, R. Andel, N. L. Pedersen, B. Johansson, and M. Gatz (2003). "Does participation in leisure activities lead to reduced risk of Alzheimer's disease? A prospective study of Swedish twins," *The Journals of Gerontology Series B: Psychological Sciences and Social Sciences*, 58(5), pp. P249–P255. DOI: 10.1093/geronb/58.5.P249. 90

[156]   A. Passmore (2003). "The occupation of leisure: Three typologies and their influence on mental health in adolescence," *OTJR: Occupation, Participation and Health*, 23(2), pp. 76–83. DOI: 10.1177/153944920302300205. 90

[157]   G. N. Siperstein, G. C. Glick, and R. C. Parker (2009). "Social inclusion of children with intellectual disabilities in a recreational setting," *Intellectual and Developmental Disabilities*, 47(2), pp. 97–107. DOI: 10.1352/1934-9556-47.2.97. 90

[158] W.-Y. G. Louie and G. Nejat (2016). "A learning from demonstration system architecture for robots learning social group recreational activities," in *RSJ International Conference on Intelligent Robots and Systems (IROS)*, pp. 808–814: IEEE. DOI: 10.1109/IROS.2016.7759144. 90

[159] D. Cox. (November 29, 2017). "Can these little robots ease the big eldercare crunch?" Available: https://www.nbcnews.com/mach/science/can-these-little-robots-ease-big-eldercare-crunch-ncna819841. 92

[160] J. Li, W.-Y. G. Louie, F. Despond, and G. Nejat (2016). "A user-study with tangy the bingo facilitating robot and long-term care residents," in *Proceedings of the IEEE International Symposium on Robotics and Intelligent Sensors (IRIS)*, December: IEEE. DOI: 10.1109/IRIS.2016.8066075. 90, 91

[161] C. Thompson, S. Mohamed, W.-Y. G. Louie, J. C. He, J. Li, and G. Nejat (2017). "The robot Tangy facilitating trivia games: A team-based user-study with long-term care residents," *IEEE 5th International Symposium on Robotics and Intelligent Sensors (IRIS 2017)*, October, pp. 173–178. DOI: 10.1109/IRIS.2017.8250117. 90

[162] K. E. C. Booth, S. C. Mohamed, S. Rajaratnam, G. Nejat, and J. C. Beck (2017). "Robots in retirement homes: Person search and task planning for a group of residents by a team of assistive robots," *IEEE Intelligent Systems*, 32(6), pp.14–21. DOI: 10.1109/MIS.2017.4531227. 90

[163] S. C. Mohamed and G. Nejat (2016). "Autonomous search by a socially assistive robot in a residential care environment for multiple elderly users using group activity preferences," in *Proceedings of the 26th International Conference on Automated Planning and Scheduling (ICAPS) Workshop on Planning and Robotics (PlanRob)*, pp. 58–66. 90

[164] P. Bovbel and G. Nejat (2014). "Casper: An assistive kitchen robot to promote aging in place," *Journal of Medical Devices - Transactions of the ASME*, 8(3), Art. no. 030945. DOI: 10.1115/1.4027113. 92

[165] L. Woiceshyn, Y. Wang, G. Nejat, and B. Benhabib (2017). "Personalized clothing recommendation by a social robot," *IEEE 5th International Symposium on Robotics and Intelligent Sensors (IRIS 2017)*, October, pp.179–185. DOI: 10.1109/IRIS.2017.8250118. 92

[166] S. Herath, M. Harandi, and F. Porikli (2017). "Going deeper into action recognition: A survey," *Image and Vision Computing*, 60, pp. 4–21, 2017/04/01/.

[167] Max Planck Institute for Informatics. (2014, FEb 4). "MPII cooking activities dataset." Available: https://www.mpi-inf.mpg.de/departments/computer-vision-and-multimodal-computing/research/human-activity-recognition/mpii-cooking-activities-dataset/.

[168]   National Institute for Health Research. (February 4, 2018). "What is public involvement in research?" Available: http://www.invo.org.uk/find-out-more/what-is-public-involvement-in-research-2/.

[169]   R. Levinson (1995). "A general programming language for unified planning and control," *Artificial Intelligence: Special Issue on Planning and Scheduling*, 76, July.

[170]   P. Viswanathan, R. C. Simpson, G. Foley, A. Sutcliffe, and J. Bell (2017). "Smart wheelchairs for assessment and mobility" (Book chapter). *Robotic Assistive Technologies: Principles and Practice*. Eds. P. Encarnação, A. Cook. CRC Press, Taylor and Francis Group. DOI: 10.1201/9781315368788-6. 82

[171]   P. Viswanathan, E. P. Zambalde, G. Foley, J. Bell, R. H. Wang, B. Adhikari, A. Mackworth, A. Mihailidis, W. C. Miller, and I. M. Mitchell (2017). "Intelligent wheelchair control strategies for older adults with cognitive impairment: User attitudes, needs, and preferences," *Autonomous Robots*, 41(3), pp. 539–554. DOI: 10.1007/s10514-016-9568-y. 82

[172]   P. W. Rushton, B. W. Mortenson, P. Viswanathan, R.H. Wang, W.C. Miller, and L. Hurd Clarke (2016). "Intelligent power wheelchair use in long-term care: potential users' experiences and perceptions," *Disability and Rehabilitation: Assistive Technology*, pp. 1–7. DOI: 10.1080/17483107.2016.1260653. 82

[173]   D. Schacter, C. Wang, G. Nejat, and B. Benhabib (2013). "A two-dimensional facial-affect estimation system for human-robot interaction using facial expression parameters," *Advanced Robotics Journal*, 27(4), pp. 259–273. DOI: 10.1080/01691864.2013.755278. 90

[174]   D. McColl and G. Nejat (2014). "Determining the affective body language of older adults during socially assistive HRI," *IEEE International Conference on Intelligent Robots and Systems*, pp. 2633–2638. DOI: 10.1109/IROS.2014.6942922. 90

Author Biographies

Dr. Jennifer Boger has been an active member in the field of intelligent assistive technology for enhancing safety and independence for older adults and people with disabilities for more than 15 years. She is the Schlegel Chair in Technology for Independent Living and an Assistant Professor in the Department of Systems Design Engineering at the University of Waterloo, with cross-appointments to the School of Public Health and Health Sciences, University of Waterloo and Affiliate Scientist at the Toronto Rehabilitation Institute. Jennifer's goal is to create innovative, accessible technologies that simultaneously enrich the experience of aging as well as society's perception of aging. She has published more than 50 peer-reviewed journals and conference contributions and is an active member of the rehabilitation engineering community.

Dr. Victoria Young is a professional engineer specializing in systems and clinical engineering and has been conducting research for 14 years in the field of rehabilitation, specifically assistive and medical technologies. Her research largely focuses on examining how individuals with complex care needs, especially older adults, access medical care at home and in the community, and how technologies can be better integrated into this care process. She has published work that explores ways in which one might enhance the robustness of spoken dialogue-based systems within artificially intelligent, personal emergency response systems. She is also investigating how medical diagnostic services could be better coordinated to provide urgent care at home for high cost hospital users needing more timely access to medical care.

Dr. Jesse Hoey is an assistant professor in the David R. Cheriton School of Computer Science at the University of Waterloo. He is also an adjunct scientist at the Toronto Rehabilitation Institute in Toronto, Canada and an Honorary Research Fellow at the University of Dundee, Scotland. His research focuses on planning and acting in large-scale real-world uncertain domains. He has published over 30 peer-reviewed scientific papers in highly visible journals and conferences. He won the Microsoft/AAAI Distinguished Contribution Award at the 2009 IJCAI Workshop on Intelligent Systems for Assisted Cognition for his paper on technology to facilitate creative expression in persons with dementia. He won the Best Paper award at the International Conference on Vision Systems (ICVS) in 2007 for his paper describing an assistive system during hand washing for persons with dementia. Dr. Hoey is program chair of the 2011 British Machine Vision Conference (BMVC).

Dr. Tizneem Jiancaro is a design researcher and writer. Her primary interest lies in developing accessible mainstream technologies for people with cognitive impairments. She is also a keen advocate of empathy-based design practice, having written in depth on this topic. Her background includes a dual Ph.D. in Biomedical Engineering and Rehabilitation Science, and undergraduate degrees in Mechanical Engineering and Cognitive Science.

Dr. Alex Mihailidis is the Barbara G. Stymiest Research Chair in Rehabilitation Technology at the University of Toronto and Toronto Rehab Institute. He is a Professor in the Department of Occupational Science and Occupational Therapy and in the Institute of Biomaterials and Biomedical Engineering, with a cross appointment in the Department of Computer Science (all at the University of Toronto). He has been conducting research in the field of pervasive computing and intelligent systems in health for the past 15 years, having published over 150 journal papers, conference papers, and abstracts in this field. He has specifically focused on the development of intelligent home systems for elder care and wellness, technology for children with autism, and adaptive tools for nurses and clinical applications.

Printed in the United States
by Baker & Taylor Publisher Services